The Coordinate-Free Approach to Linear Models

This book is about the coordinate-free, or geometric, approach to the theory of linear models, more precisely, Model I ANOVA and linear regression models with nonrandom predictors in a finite-dimensional setting. This approach is more insightful, more elegant, more direct, and simpler than the more common matrix approach to linear regression, analysis of variance, and analysis of covariance models in statistics. The book discusses the intuition behind and optimal properties of various methods of estimating and testing hypotheses about unknown parameters in the models.

Topics covered include inner product spaces, orthogonal projections, book orthogonal spaces, Tjur experimental designs, basic distribution theory, the geometric version of the Gauss-Markov theorem, optimal and nonoptimal properties of Gauss-Markov, Bayes and shrinkage estimators under the assumption of normality, the optimal properties of F-tests, and the analysis of covariance and missing observations.

Michael J. Wichura has 37 years of teaching experience in the Department of Statistics at the University of Chicago. He has served as an associate editor for the *Annals of Probability* and was the database editor for the *Current Index to Statistics* from 1995 to 2000. He is the author of the PiCTeX macros (for drawing pictures in TeX) and the PiCTeX manual and also of the TABLE macros and the TABLE manual.

The Coordinate-Free Approach
to Linear Models

MICHAEL J. WICHURA

CAMBRIDGE
UNIVERSITY PRESS

CAMBRIDGE
UNIVERSITY PRESS

32 Avenue of the Americas, New York NY 10013-2473, USA

Cambridge University Press is part of the University of Cambridge.

It furthers the University's mission by disseminating knowledge in the pursuit of education, learning and research at the highest international levels of excellence.

www.cambridge.org
Information on this title: www.cambridge.org/9780521868426

First published 2006

A catalogue record for this publication is available from the British Library

Library of Congress Cataloguing in Publication data

Wichura, Michael J. (Michael John)
The coordinate-free approach to linear models / Michael J. Wichura.
    p. cm. – (Cambridge series on statistical and probabilistic mathematics)
Includes bibliographical references and index.
ISBN-13: 978-0-521-86842-6 (hardback)
ISBN-10: 0-521-86842-4 (hardback)
1. Linear models (Statistics)   2. Analysis of variance.   3. Regression analysis.
4. Analysis of covariance.   I. Title.   II. Series.
QA279.W53 2006
519.5–dc22   2006004133

ISBN 978-0-521-86842-6 Hardback

In memoriam
William H. Kruskal
1919–2005

CONTENTS

PREFACE

When I was a graduate student in the mid-1960s, the mathematical theory underlying analysis of variance and regression became clear to me after I read a draft of William Kruskal's monograph on the so-called coordinate-free, or geometric, approach to these subjects. Alas, with Kruskal's demise, this excellent treatise will never be published.

From time to time during the 1970s, 80s, and early 90s, I had the good fortune to teach the coordinate-free approach to linear models, more precisely, to Model I analysis of variance and linear regression with nonrandom predictors. While doing so, I evolved my own set of lecture notes, presented here. With regard to inspiration and content, my debt to Kruskal is clear. However, my notes differ from Kruskal's in many ways. To mention just a few, my notes are intended for a one- rather than three-quarter course. The notes are aimed at statistics graduate students who are already familiar with the basic concepts of linear algebra, such as linear subspaces and linear transformations, and who have already had some exposure to the matricial formulation of the GLM, perhaps through a methodology course, and who are interested in the underlying theory. I have also included Tjur experimental designs and some of the highlights of the optimality theory for estimation and testing in linear models under the assumption of normality, feeling that the elegant setting provided by the coordinate-free approach is a natural one in which to place these jewels of mathematical statistics. As he alluded to in his conversation with Zabell (1994), Kruskal always wished that he could have brought "his book, his potential book, his unborn book" to life. Out of deference to Kruskal, who was my colleague here at the University of Chicago, I have not until now made my notes public.

For motivation, Chapter 1 presents an example illustrating Kruskal's claim in his 1961 Berkeley Symposium paper that "the coordinate-free approach ... permits a simpler, more general, more elegant, and more direct treatment of the general theory ... than do its notational counterparts, the matrix and scalar approaches." I hope that as s/he works through the book, the reader will be more and more convinced that this is indeed the case.

The last section of Chapter 2 reviews the "elementary" concepts from linear algebra which the reader is assumed to know already. The first five sections of that chapter develop the "nonelementary" tools we need, such as (finite-dimensional, real) inner product spaces, orthogonal projections, the spectral theorem for self-adjoint linear transformations, and the representation of linear and bilinear functionals. Section 2.3 uses the notions of book orthogonal subspaces and orthogonal projections, along with the inclusion and intersection of subspaces, to discuss Tjur experimental designs, thereby giving a unified treatment of the algebra underlying the models one usually encounters in a first course in analysis of variance, and much more.

Chapter 3 develops basic distribution theory for random vectors taking values in inner product spaces — the first- and second-moment structures of such vectors, and the key fact that if one splits a spherical normal random vector up into its components in mutually orthogonal subspaces, then those components are independent and have themselves spherical normal distributions within their respective subspaces.

The geometric version of the Gauss-Markov theorem is discussed in Chapter 4, from the point of view of both estimating linear functionals of the unknown mean vector and estimating the mean vector itself. These results are based just on assumptions about the first and second moment structures of the data vector. For an especially nice example of how the geometric viewpoint is more insightful than the matricial one, be sure to work the "four-penny problem" in Exercise 4.2.17.

Estimation under the assumption of normality is taken up in Chapter 5. In some respects, Gauss-Markov estimators are optimal; for example, they are minimum variance unbiased. However, in other respects they are not optimal. Indeed, Bayesian considerations lead naturally to the James-Stein shrinkage type estimators, which can significantly outperform GMEs in terms of mean square error.

Again under the assumption of normality, F-testing of null hypotheses and the related issue of interval estimation are taken up in Chapter 6. Chapters 7 and 8 deal with the analysis of covariance and missing observations, respectively.

The book is written at the level of Halmos's *Finite-Dimensional Vector Spaces* (but Halmos is not a prerequisite). Thus the reader will on the one hand need to be comfortable with the yoga of definitions, theorems, and proofs, but on the other hand be comforted by knowing that the abstract ideas will be illustrated by concrete examples and presented with (what I hope are) some insightful comments. To get a feeling for the coordinate-free approach before embarking on a serious study of this book, you might find it helpful to first read one or more of the brief elementary nontechnical expositions of the subject that have appeared in *The American Statistician*, for example, Herr (1980), Bryant (1984), or Saville and Wood (1986). From the perspective of mathematical statistics, there are some very elegant results, and some notable surprises, connected with the optimality theory for Gauss-Markov estimation and F-testing under the assumption of normality. In the sections that deal with these matters—in particular Sections 5.4, 5.5, 6.3, 6.4, and 6.6—the mathematics is somewhat harder than elsewhere, corresponding to the greater depth of the theory.

Each of the chapters following the Introduction contains numerous exercises, along with a problem set that develops some topic complementing the material in that chapter. Altogether there are about 200 exercises. Most of them are easy but, I hope, instructive. I typically devote most of the class time to having students present solutions to the exercises. Some exercises foreshadow what is to come, by covering a special case of material that will be presented in full generality later. Moreover, the assertions of some exercises are appealed to later in the text. If you are working through the book on your own, you should at least read over each exercise, even if you do not work things out. The problem sets, several of which are based on journal articles, are harder than the exercises and require a sustained effort for their completion.

The students in my courses have typically worked through the whole book in one quarter, but that is admittedly a brisk pace. One semester would be less demanding. For a short course, you could concentrate on the parts of the book that flesh out the outline of the coordinate-free viewpoint that Kruskal set out in Section 2 of his aforementioned Berkeley Symposium paper. That would involve: Chapter 1, for motivation; Sections 2.1, 2.2, 2.5, and 2.7 for notation and basic results from linear algebra; Sections 3.1–3.8 for distribution theory; Sections 4.1–4.7 for the properties of Gauss-Markov estimators; Sections 5.1 and 5.2 for estimation under the assumption of normality; and Sections 6.1, 6.2, and 6.5 for hypothesis testing and interval estimation under normality.

Various graduate students, in particular Neal Thomas and Nathaniel Schenker, have made many comments that have greatly improved this book. My thanks go to all of them, and also to David Van Dyke and Peter Meyer for suggesting how easy/hard each of the exercises is. Thanks are also due to Mitzi Nakatsuka for her help in converting the notes to TEX, and to Persi Diaconis for his advice and encouragement.

Michael J. Wichura
University of Chicago

CHAPTER 1

INTRODUCTION

In this chapter we introduce and contrast the matricial and geometric formulations of the so-called general linear model and introduce some notational conventions.

1. Orientation

Recall the classical framework of the *general linear model (GLM)*. One is given an n-dimensional random vector $\boldsymbol{Y}^{n\times 1} = (Y_1\ldots, Y_n)^T$, perhaps multivariate normally distributed, with covariance matrix $(\mathrm{Cov}(Y_i, Y_j))^{n\times n} = \sigma^2 \boldsymbol{I}^{n\times n}$ and mean vector $\boldsymbol{\mu}^{n\times 1} = E(\boldsymbol{Y}) = (EY_1, \ldots, EY_n)^T$ of the form

$$\boldsymbol{\mu} = \boldsymbol{X}\boldsymbol{\beta},$$

where $\boldsymbol{X}^{n\times p}$ is known and σ^2 and $\boldsymbol{\beta}^{p\times 1} = (\beta_1, \ldots, \beta_p)^T$ are unknown; in addition, the β_i's may be subject to linear constraints $\boldsymbol{R}\boldsymbol{\beta} = 0$, where $\boldsymbol{R}^{c\times p}$ is known. $\boldsymbol{X}$ is called the *design*, or *regression*, *matrix*, and $\boldsymbol{\beta}$ is called the *parameter vector*.

1.1 Example. In the classical *two-sample problem*, one has

$$\boldsymbol{X}^T = \begin{pmatrix} \underbrace{1 \quad 1 \quad \ldots \quad 1}_{n_1 \text{ times}} & \underbrace{0 \quad 0 \quad \ldots \quad 0}_{} \\ 0 \quad 0 \quad \ldots \quad 0 & \underbrace{1 \quad 1 \quad \ldots \quad 1}_{n_2 \text{ times}} \end{pmatrix} \quad \text{and} \quad \boldsymbol{\beta} = \begin{pmatrix} \mu_1 \\ \mu_2 \end{pmatrix},$$

that is,

$$E(Y_i) = \begin{cases} \mu_1, & \text{if } 1 \leq i \leq n_1, \\ \mu_2, & \text{if } n_1 < i \leq n_1 + n_2 = n. \end{cases}$$

●

1.2 Example. In *simple linear regression*, one has

$$\boldsymbol{X}^T = \begin{pmatrix} 1 & 1 & \ldots & 1 \\ x_1 & x_2 & \ldots & x_n \end{pmatrix} \quad \text{and} \quad \boldsymbol{\beta} = \begin{pmatrix} a \\ b \end{pmatrix},$$

that is,

$$E(Y_i) = a + bx_i \quad \text{for } i = 1, \ldots, n.$$

●

Typical problems are the estimation of linear combinations of the β_i's, testing that some such linear combinations are 0 (or some other prescribed value), and the estimation of σ^2.

1.3 Example. In the two-sample problem, one is often interested in estimating the difference $\mu_2 - \mu_1$ or in testing the null hypothesis that $\mu_1 = \mu_2$.   •

1.4 Example. In simple linear regression, one seeks estimates of the intercept a and slope b and may want to test, for example, the hypothesis that $b = 0$ or the hypothesis that $b = 1$.   •

If you have had some prior statistical training, you may well have already encountered the resolution of these problems. You may know, for example, that provided $\boldsymbol{X}$ is of full rank and no linear constraints are imposed on $\boldsymbol{\beta}$, the *best (minimum variance) linear unbiased estimator (BLUE)* of $\sum_{1 \leq i \leq p} c_i \beta_i$ is $\sum_{1 \leq i \leq p} c_i \hat{\beta}_i$, where

$$(\hat{\beta}_1, \ldots, \hat{\beta}_p)^T = \boldsymbol{C}\boldsymbol{X}^T\boldsymbol{Y}, \quad \text{with} \quad \boldsymbol{C} = \boldsymbol{A}^{-1}, \quad \boldsymbol{A} = \boldsymbol{X}^T\boldsymbol{X};$$

this is called the *Gauss-Markov theorem*.

In this book we will be studying the GLM from a geometric point of view, using linear algebra in place of matrix algebra. Although we will not reach any conclusions that could not be obtained solely by matrix techniques, the basic ideas will emerge more clearly. With the added intuitive feeling and mathematical insight this provides, one will be better able to understand old results and formulate and prove new ones.

From a geometric perspective, the GLM may be described as follows, using some terms that will be defined in subsequent chapters. One is given a *random vector* Y taking values in some given *inner product space* $(V, \langle \cdot, \cdot \rangle)$. It is assumed that Y has a *weakly spherical covariance operator* and the *mean* μ of Y lies in a given *manifold* M of V; for purposes of testing, it is further assumed that Y is *normally distributed*. One desires to estimate μ (or *linear functionals* of μ) and to test hypotheses such as $\mu \in M_0$, where M_0 is a given *submanifold* of M. The Gauss-Markov theorem says that the BLUE of the linear functional $\psi(\mu)$ is $\psi(\hat{\mu})$, where $\hat{\mu}$ is the *orthogonal projection* of Y onto M. As we will see, this geometric description of the problem encompasses the matricial formulation of the GLM not only as it is set out above (take, for example, $V = \mathbb{R}^n$, $\langle \cdot, \cdot \rangle = $ dot-product, $Y = \boldsymbol{Y}$, $\mu = \boldsymbol{\mu}$, and $M = $ the subspace of $\mathbb{R}^n$ spanned by the columns of the design matrix $\boldsymbol{X}$), but also in cases where $\boldsymbol{X}$ is of less than full rank and/or linear constraints are imposed on the β_i's.

2. An illustrative example

To illustrate the differences between the matricial and geometric approaches, we compare the ways in which one establishes the independence of

$$\bar{Y} = \hat{\mu} = \frac{\sum_{1 \leq i \leq n} Y_i}{n} \quad \text{and} \quad s^2 = \hat{\sigma}^2 = \frac{1}{n-1} \sum_{1 \leq i \leq n} (Y_i - \bar{Y})^2$$

in the one-sample problem

$$Y^{n \times 1} \sim N(\mu e, \sigma^2 I^{n \times n}) \quad \text{with} \quad e = (1, 1, \ldots, 1)^T. \tag{2.1}$$

(The vector e is called the *equiangular vector*.)

The classical matrix proof, which uses some facts about multivariate normal distributions, runs like this. Let $B^{n \times n} = (b_{ij})$ be the matrix

$$\begin{pmatrix} \frac{1}{\sqrt{n}} & \frac{1}{\sqrt{n}} & \frac{1}{\sqrt{n}} & \frac{1}{\sqrt{n}} & \cdots & \frac{1}{\sqrt{n}} & \frac{1}{\sqrt{n}} \\[2mm] \frac{1}{\sqrt{2}} & \frac{-1}{\sqrt{2}} & 0 & 0 & \cdots & 0 & 0 \\[2mm] \frac{1}{\sqrt{6}} & \frac{1}{\sqrt{6}} & \frac{-2}{\sqrt{6}} & 0 & \cdots & 0 & 0 \\[2mm] \frac{1}{\sqrt{12}} & \frac{1}{\sqrt{12}} & \frac{1}{\sqrt{12}} & \frac{-3}{\sqrt{12}} & \cdots & 0 & 0 \\[2mm] \vdots & \vdots & \vdots & \vdots & \cdots & \vdots & \vdots \\[2mm] \frac{1}{\sqrt{n(n-1)}} & \frac{1}{\sqrt{n(n-1)}} & \frac{1}{\sqrt{n(n-1)}} & \frac{1}{\sqrt{n(n-1)}} & \cdots & \frac{1}{\sqrt{n(n-1)}} & \frac{-(n-1)}{\sqrt{n(n-1)}} \end{pmatrix}.$$

Note that the rows (and columns) of B are orthonormal $\left(\sum_{1 \le k \le n} b_{ik} b_{jk} = \delta_{ij} \equiv \begin{cases} 1, & \text{if } i = j \\ 0, & \text{if } i \ne j \end{cases} \right)$ and that the first row is $\frac{1}{\sqrt{n}} e^T$. Set

$$Z = BY.$$

Then

$$Z \sim N(\nu, \Sigma)$$

with

$$\nu = B(\mu e) = \mu B e = (\sqrt{n}\,\mu, 0, \ldots, 0)^T$$

and

$$\Sigma = B(\sigma^2 I) B^T = \sigma^2 B B^T = \sigma^2 I;$$

that is, $Z_1, Z_2, \ldots, Z_n$ are independent normal random variables, each with variance σ^2, $E(Z_1) = \sqrt{n}\,\mu$, and $E(Z_j) = 0$ for $2 \le j \le n$. Moreover,

$$Z_1 = \frac{1}{\sqrt{n}} \sum_{1 \le i \le n} Y_i = \sqrt{n}\,\bar{Y}, \text{ or } \bar{Y} = \frac{Z_1}{\sqrt{n}},$$

while

$$\begin{aligned} (n-1)s^2 &= \sum_{1 \le i \le n} (Y_i - \bar{Y})^2 = \sum_{1 \le i \le n} Y_i^2 - n\bar{Y}^2 \\ &= \sum_{1 \le i \le n} Y_i^2 - Z_1^2 = \sum_{1 \le i \le n} Z_i^2 - Z_1^2 = \sum_{2 \le i \le n} Z_i^2 \end{aligned} \tag{2.2}$$

because

$$\sum_{1 \leq i \leq n} Z_i^2 = \boldsymbol{Z}^T \boldsymbol{Z} = \boldsymbol{Y}^T \boldsymbol{B}^T \boldsymbol{B} \boldsymbol{Y} = \boldsymbol{Y}^T \boldsymbol{Y} = \sum_{1 \leq i \leq n} Y_i^2.$$

This gives the independence of $\bar{Y}$ and s^2, and it is an easy step to get the marginal distributions: $\bar{Y} \sim N(\mu, \sigma^2/n)$ and $(n-1)s^2/\sigma^2 \sim \chi_{n-1}^2$.

What is the nature of the transformation $\boldsymbol{Z} = \boldsymbol{BY}$? Let $\boldsymbol{b}_1 = e/\sqrt{n}$, $\boldsymbol{b}_2, \ldots, \boldsymbol{b}_n$ denote the transposes of the rows of $\boldsymbol{B}$. The coordinates of $\boldsymbol{Y} = \sum_{1 \leq j \leq n} C_j \boldsymbol{b}_j$ with respect to this new orthonormal basis for $\mathbb{R}^n$ are given by

$$C_i = \boldsymbol{b}_i^T \boldsymbol{Y} = Z_i, \quad i = 1, \ldots, n.$$

The effect of the change of coordinates $\boldsymbol{Y} \to \boldsymbol{Z}$ is to split $\boldsymbol{Y}$ into its components along, and orthogonal to, the equiangular vector e.

Now I will show you the geometric proof, which uses some properties of (weakly) spherical normal random vectors taking values in an inner product space $(V, \langle \cdot, \cdot \rangle)$, here $(\mathbb{R}^n, \text{dot-product})$. The assumptions imply that $\boldsymbol{Y}$ is spherical normally distributed about its mean $E(\boldsymbol{Y})$ and $E(\boldsymbol{Y})$ lies in the manifold M spanned by e. Let P_M denote orthogonal projection onto M and Q_M orthogonal projection onto the orthogonal complement $M^\perp$ of M. Basic distribution theory says that $P_M \boldsymbol{Y}$ and $Q_M \boldsymbol{Y}$ are independent. But

$$P_M \boldsymbol{Y} = \frac{\langle e, \boldsymbol{Y} \rangle}{\langle e, e \rangle} e = \bar{Y} e \tag{2.3}$$

and

$$Q_M \boldsymbol{Y} = \boldsymbol{Y} - P_M \boldsymbol{Y} = \boldsymbol{Y} - \bar{Y} e = (Y_1 - \bar{Y}, \ldots, Y_n - \bar{Y})^T;$$

it follows that $\bar{Y}$ and $(n-1)s^2 = \sum_{1 \leq i \leq n}(Y_i - \bar{Y})^2 = \|Q_M \boldsymbol{Y}\|^2$ are independent. Again it is an easy matter to get the marginal distributions.

To my way of thinking, granted the technical apparatus, the second proof is clearer, being more to the point. The first proof does the same things, but (to the uninitiated) in an obscure manner.

3. Notational conventions

The chapters are organized into sections. Within each section of the current chapter, enumerated items are numbered consecutively in the form

$$(\mathit{section\_number.item\_number}).$$

References to items in a different chapter take the expanded form

$$(\mathit{chapter\_number.section\_number.item\_number}).$$

For example, (2.4) refers to the 4th numbered item (which may be an example, exercise, theorem, formula, or whatever) in the 2nd section of the current chapter, while (6.1.3) refers to the 3rd numbered item in the 1st section of the 6th chapter.

Each exercise is assigned a difficulty level using the syntax

Exercise [d],

where d is an integer in the range 1 to 5 — the larger is d, the harder the exercise. The value of d depends both on the intrinsic difficulty of the exercise and the length of time needed to write up the solution.

To help distinguish between the matricial and geometric points of view, matrices, including row and column vectors, are written in ***italic boldface*** type while linear transformations and elements of abstract vector spaces are written simply in *italic* type. We speak, for example, of the design matrix $\boldsymbol{X}$ but of vectors v and w in an inner product space V.

The end of a proof is marked by a ∎, of an example by a •, of an exercise by a ◇, and of a part of problem set by a ∘.

CHAPTER 2

TOPICS IN LINEAR ALGEBRA

In this chapter we discuss some topics from linear algebra that play a central role in the geometrical analysis of the GLM. The notion of orthogonal projection in an inner product space is introduced in Section 2.1 and studied in Section 2.2. A class of orthogonal decompositions that are useful in the design of experiments is studied in Section 2.3. The spectral representation of self-adjoint transformations is developed in Section 2.4. Linear and bilinear functionals are discussed in Section 2.5. The chapter closes with a problem set in Section 2.6, followed by an appendix containing a brief review of the basic definitions and facts from linear algebra with which we presume the reader is already familiar.

1. Orthogonal projections

Throughout this book we operate in the context of a finite-dimensional *inner product space* $(V, \langle \cdot, \cdot \rangle)$ — V is a finite-dimensional vector space and $\langle \cdot, \cdot \rangle \colon V \times V \to \mathbb{R}$ is an *inner product:*

(i) *(positive-definiteness)* $\langle x, x \rangle \geq 0$ for all $x \in V$ and $\langle x, x \rangle = 0$ if and only if $x = 0$.

(ii) *(symmetry)* $\langle x, y \rangle = \langle y, x \rangle$ for all $x, y \in V$.

(iii) *(bilinearity)* For all $c_1, c_2 \in \mathbb{R}$ and $x_1, x_2, x, y_1, y_2, y \in V$, one has

$$\langle c_1 x_1 + c_2 x_2, y \rangle = c_1 \langle x_1, y \rangle + c_2 \langle x_2, y \rangle$$
$$\langle x, c_1 y_1 + c_2 y_2 \rangle = c_1 \langle x, y_1 \rangle + c_2 \langle x, y_2 \rangle.$$

The canonical example is $V = \mathbb{R}^n$ endowed with the dot-product

$$\langle \boldsymbol{x}, \boldsymbol{y} \rangle = \boldsymbol{x} \cdot \boldsymbol{y} = \sum_{1 \leq i \leq n} x_i y_i = \boldsymbol{x}^T \boldsymbol{y}$$

for $\boldsymbol{x} = (x_1, \ldots, x_n)^T$ and $\boldsymbol{y} = (y_1, \ldots, y_n)^T$. Unless specifically stated to the contrary, we always view $\mathbb{R}^n$ as endowed with the dot-product.

Two vectors x and y in V are said to be *perpendicular*, or *orthogonal* (with respect to $\langle \cdot, \cdot \rangle$), if

$$\langle x, y \rangle = 0;$$

one writes

$$x \perp y.$$

The quantity

$$\|x\| = \sqrt{\langle x, x \rangle}$$

is called the *length*, or *norm*, of x. The squared length of the sum of two vectors is given by

$$\|x + y\|^2 = \langle x + y, x + y \rangle = \langle x, x + y \rangle + \langle y, x + y \rangle$$
$$= \|x\|^2 + 2\langle x, y \rangle + \|y\|^2,$$

which reduces to the *Pythagorean theorem*,

$$\|x + y\|^2 = \|x\|^2 + \|y\|^2, \tag{1.1}$$

when $x \perp y$.

1.2 Exercise [1]. Let v_1 and v_2 be two nonzero vectors in $\mathbb{R}^2$ and let θ be the angle between them, measured counter-clockwise from v_1 to v_2. Show that

$$\cos(\theta) = \frac{\langle v_1, v_2 \rangle}{\|v_1\| \, \|v_2\|}$$

and deduce that $v_1 \perp v_2$ if and only if $\theta = 90°$ or $270°$.
[Hint: Use the identity $\cos(\theta_2 - \theta_1) = \cos(\theta_1)\cos(\theta_2) + \sin(\theta_1)\sin(\theta_2)$.] ◇

1.3 Exercise [3]. Let x, y, and z be the vectors in $\mathbb{R}^n$ given by

$$x_i = 1, \ y_i = i - \frac{n+1}{2}, \ \text{and} \ z_i = \left(i - \frac{n+1}{2}\right)^2 - \frac{n^2-1}{12} \quad \text{for } 1 \leq i \leq n. \tag{1.4}$$

Show that x, y, and z are mutually orthogonal and span the same subspace as do $(1, 1, \ldots, 1)^T$, $(1, 2, \ldots, n)^T$, and $(1, 4, \ldots, n^2)^T$. Exhibit x, y, and z explicitly for $n = 5$. ◇

1.5 Exercise [1]. Let $\mathcal{O} \colon V \to V$ be a linear transformation. Show that $\mathcal{O}$ preserves lengths, that is,

$$\|\mathcal{O}x\| = \|x\| \quad \text{for all } x \in V$$

if and only if it preserves inner products, that is,

$$\langle \mathcal{O}x, \mathcal{O}y \rangle = \langle x, y \rangle \quad \text{for all } x, y \in V.$$

Such a transformation is said to be *orthogonal*.
[Hint: Observe that

$$\langle x, y \rangle = \frac{\|x + y\|^2 - \|x\|^2 - \|y\|^2}{2} \tag{1.6}$$

for $x, y \in V$.] ◇

1.7 Exercise [2]. (1) Let v_1 and v_2 be elements of V. Show that $v_1 = v_2$ if and only if $\langle v_1, w \rangle = \langle v_2, w \rangle$ for each $w \in V$, or just for each w in some basis for V. (2) Let T_1 and T_2 be two linear transformations of V. Show that $T_1 = T_2$ if and only if $\langle v, T_1 w \rangle = \langle v, T_2 w \rangle$ for each v and w in V, or just for each v and w in some basis for V. ◇

As intimated in the Introduction, the notion of orthogonal projection onto subspaces of V plays a key role in the study of the GLM. We begin our study of projections with the following seminal result.

1.8 Theorem. *Suppose that M is a subspace of V and that $x \in V$. There is exactly one vector $m \in M$ such that the residual $x - m$ is orthogonal to M:*

$$(x - m) \perp y \quad \text{for all } y \in M, \tag{1.9}$$

or, equivalently, such that m is closest to x:

$$\|x - m\| = \inf\{\, \|x - y\| : y \in M \,\}. \tag{1.10}$$

The proof will be given shortly. The unique $m \in M$ such that (1.9) and (1.10) hold is called the *orthogonal projection of x onto M*, written $P_M x$, and the mapping P_M that sends $x \in V$ to $P_M x$ is called *orthogonal projection onto M*. In the context of $\mathbb{R}^n$ with $\boldsymbol{x} = (x_i)$ and $\boldsymbol{m} = (m_i)$, (1.9) reads

$$\sum_{1 \leq i \leq n} (x_i - m_i) y_i = 0 \quad \text{for all } \boldsymbol{y} = (y_i) \in M,$$

while (1.10) is the *least squares characterization* of $\boldsymbol{m}$:

$$\sum_{1 \leq i \leq n} (x_i - m_i)^2 = \inf\Big\{ \sum_{1 \leq i \leq n} (x_i - y_i)^2 : \boldsymbol{y} \in M \Big\}.$$

1.11 Exercise [1]. Let M be a subspace of V. (a) Show that

$$\|x - P_M x\|^2 = \|x\|^2 - \|P_M x\|^2 \tag{1.12}$$

for all $x \in V$. (b) Deduce that

$$\|P_M x\| \leq \|x\| \tag{1.13}$$

for all $x \in V$, with equality holding if and only if $x \in M$. ◇

Proof of Theorem 1.8. (1.9) *implies* (1.10): Suppose $m \in M$ satisfies (1.9). Then for all $y \in M$ the Pythagorean theorem gives

$$\|x - y\|^2 = \|(x - m) + (m - y)\|^2 = \|x - m\|^2 + \|m - y\|^2,$$

so m satisfies (1.10).

(1.10) *implies* (1.9): Suppose $m \in M$ satisfies (1.10). Then, for any $0 \neq y \in M$ and any $\delta \in \mathbb{R}$,

$$\|x - m\|^2 \leq \|x - m + \delta y\|^2 = \|x - m\|^2 + 2\delta \langle x - m, y \rangle + \delta^2 \|y\|^2;$$

this relation forces $\langle x - m, y \rangle = 0$.

Uniqueness: If

$$m_1 + y_1 = x = m_2 + y_2$$

with $m_i \in M$ and $y_i \perp M$ for $i = 1$ and 2, then the vector

$$m_1 - m_2 = y_2 - y_1$$

lies in M and is perpendicular to M, and so it is perpendicular to itself:

$$0 = \langle m_1 - m_2, m_1 - m_2 \rangle = \|m_1 - m_2\|^2,$$

whence $m_1 = m_2$ by the positive-definiteness of $\langle \cdot, \cdot \rangle$.

Existence: We will show in a moment that M has a basis $m_1, \ldots, m_k$ consisting of mutually orthogonal vectors. For any such basis the generic $m = \sum_{1 \leq j \leq k} c_j m_j$ in M satisfies

$$(x - m) \perp M$$

if and only if $x - m$ is orthogonal to each m_i, that is, if and only if

$$\langle m_i, x \rangle = \left\langle m_i, \sum_{1 \leq j \leq k} c_j m_j \right\rangle = \sum_{1 \leq j \leq k} \langle m_i, m_j \rangle c_j = \langle m_i, m_i \rangle c_i$$

for $1 \leq i \leq k$. It follows that we can take

$$P_M x = \sum_{1 \leq i \leq k} \frac{\langle m_i, x \rangle}{\langle m_i, m_i \rangle} \, m_i. \qquad (1.14)$$

To produce an orthogonal basis for M, let $m_1^*, \ldots, m_k^*$ be any basis for M and inductively define new basis vectors $m_1, \ldots, m_k$ by the recipe $m_1 = m_1^*$ and

$$m_j = m_j^* - P_{[m_1^*, \ldots, m_{j-1}^*]} m_j^* = m_j^* - P_{[m_1, \ldots, m_{j-1}]} m_j^*$$

$$= m_j^* - \sum_{1 \leq i \leq j-1} \frac{\langle m_j^*, m_i \rangle}{\langle m_i, m_i \rangle} \, m_i \qquad (1.15)$$

for $j = 2, \ldots, k$; here $[m_1^*, \ldots, m_{j-1}^*]$ denotes the span of $m_1^*, \ldots, m_{j-1}^*$ and $[m_1, \ldots, m_{j-1}]$ denotes the (identical) span of $m_1, \ldots, m_{j-1}$ (see Subsection 2.6.1). ∎

The recursive scheme for cranking out the m_j's above is called *Gram-Schmidt orthogonalization*. As a special case of (1.14) we have the following simple, yet key, formula for projecting onto a one-dimensional space:

$$P_{[m]} x = \frac{\langle x, m \rangle}{\langle m, m \rangle} \, m \qquad \text{for } m \neq 0. \qquad (1.16)$$

1.17 Example. In the context of $\mathbb{R}^n$ take $\boldsymbol{m} = e \equiv (1, \ldots, 1)^T$. Formula (1.16) then reads

$$P_{[e]} \boldsymbol{x} = \frac{\langle \boldsymbol{x}, e \rangle}{\langle e, e \rangle} \, e = \frac{\sum_i x_i}{\sum_i 1} \, e = \bar{x} e = (\bar{x}, \ldots, \bar{x})^T;$$

we used this result in the Introduction (see (1.2.3)). Formula (1.12) reads

$$\sum_i (x_i - \bar{x})^2 = \|\boldsymbol{x} - \bar{x}\boldsymbol{e}\|^2 = \|\boldsymbol{x}\|^2 - \|\bar{x}\boldsymbol{e}\|^2 = \sum_i x_i^2 - n\bar{x}^2;$$

this is just the computing formula for $(n-1)s^2$ used in (1.2.2). According to (1.10), $c = \bar{x}$ minimizes the sum of squares $\sum_i (x_i - c)^2$.   •

1.18 Exercise [2]. Use the preceding techniques to compute $P_M \boldsymbol{y}$ for $\boldsymbol{y} \in \mathbb{R}^n$ and $M = [\boldsymbol{e}, (x_1, \ldots, x_n)^T]$, the manifold spanned by the columns of the design matrix for simple linear regression.   ◇

1.19 Exercise [3]. Show that the transposes of the rows of the matrix $\boldsymbol{B}$ of Section 1.2 result from first applying the Gram-Schmidt orthogonalization scheme to the vectors $\boldsymbol{e}$, $(1, -1, 0, \ldots, 0)^T$, $(0, 1, -1, 0, \ldots, 0)^T$, $\ldots$, $(0, \ldots, 0, 1, -1)^T$ in $\mathbb{R}^n$ and then normalizing to unit length.   ◇

1.20 Exercise [2]. Suppose $x, y \in V$. Prove the *Cauchy-Schwarz inequality*:

$$|\langle x, y \rangle| \leq \|x\| \, \|y\|, \tag{1.21}$$

with equality if and only if x and y are linearly dependent. Deduce *Minkowski's inequality*:

$$\|x + y\| \leq \|x\| + \|y\|. \tag{1.22}$$

[Hint: For (1.21), take $m = y$ in (1.16) and use part (b) of Exercise 1.11.]   ◇

1.23 Exercise [4]. Define $d: V \times V \to \mathbb{R}$ by $d(x, y) = \|y - x\|$. Show that d is a metric on V such that the set $\{\, x \in V : \|x\| = 1 \,\}$ is compact.   ◇

1.24 Exercise [3]. Show that for any two subspaces M and N of V,

$$\sup\{\, \|P_M x\| : x \in N \text{ and } \|x\| = 1 \,\} \leq 1, \tag{1.25}$$

with equality holding if and only if M and N have a nonzero vector in common.
[Hint: A continuous real-valued function on a compact set attains its maximum.]   ◇

To close this section we generalize (1.14) to cover the case of an arbitrary basis for M. Suppose then that the basis vectors $m_1, \ldots, m_k$ are not necessarily orthogonal and let $x \in V$. As in the derivation of (1.14),

$$P_M x = \sum_j c_j m_j$$

is determined by the condition

$$m_i \perp (x - P_M x) \quad \text{for } i = 1, \ldots, k,$$

that is, by the so-called *normal equations*

$$\sum_{1 \leq j \leq k} \langle m_i, m_j \rangle c_j = \langle m_i, x \rangle, \quad i = 1, \ldots, k. \tag{1.26}$$

In matrix notation (1.26) reads

$$\langle m, m \rangle c = \langle m, x \rangle,$$

so that

$$c = \langle m, m \rangle^{-1} \langle m, x \rangle,$$

the $k \times k$ matrix $\langle m, m \rangle$ and the $k \times 1$ column vectors $\langle m, x \rangle$ and c being given by

$$\left(\langle m, m \rangle \right)_{ij} = \langle m_i, m_j \rangle, \quad \left(\langle m, x \rangle \right)_i = \langle m_i, x \rangle, \quad \text{and} \quad (c)_i = c_i$$

for $i, j = 1, \ldots, k$. If the m_i's are mutually orthogonal, $\langle m, m \rangle$ is diagonal and (1.14) follows immediately.

1.27 Exercise [3]. Show that the matrix $\langle m, m \rangle$ above is in fact nonsingular, so that the inverse $\langle m, m \rangle^{-1}$ exists.
[Hint: Show that $\sum_j \langle m_i, m_j \rangle c_j = 0$ for each i implies $c_1 = \cdots = c_k = 0$.] ◇

1.28 Exercise [2]. Redo Exercise 1.18 using (1.26). Check that the two ways of calculating $P_M y$ do in fact give the same result. ◇

1.29 Exercise [2]. Let M be the subspace of $\mathbb{R}^n$ spanned by the columns of an $n \times k$ matrix X. Supposing these columns are linearly independent, use (1.26) to show that the matrix representing P_M with respect to the usual coordinate basis of $\mathbb{R}^n$ is $X(X^T X)^{-1} X^T$. Exhibit this matrix explicitly in the case $X = e$. ◇

1.30 Exercise [4]. Suppose $X_0, X_1, \ldots, X_k$ are square integrable random variables defined on a common probability space. Consider using $X_1, \ldots, X_k$ to predict X_0. Show that among predictors of the form

$$\hat{X}_0 = c_0 + \sum_{1 \leq i \leq k} c_i X_i,$$

with the c's being constants, the one minimizing the *mean square error of prediction*

$$E(\hat{X}_0 - X_0)^2$$

is

$$\hat{X}_0 = \mu_0 + \sum_{1 \leq i \leq k} \left(\sum_{1 \leq j \leq k} \sigma^{ij} \sigma_{j0} \right)(X_i - \mu_i),$$

where

$$\mu_i = E(X_i), \qquad 0 \leq i \leq k,$$
$$\sigma_{ij} = \mathrm{Cov}(X_i, X_j), \qquad 0 \leq i, j \leq k,$$

and the $k \times k$ matrix $(\sigma^{ij})_{1 \leq i, j \leq k}$ is the inverse of the matrix $(\sigma_{ij})_{1 \leq i, j \leq k}$, with the latter assumed to be nonsingular.
[Hint: This is just a matter of projecting $X_0 - \mu_0$ onto the subspace spanned by 1, $X_1 - \mu_1, \ldots, X_k - \mu_k$ in the space $\mathcal{L}_2$ of square integrable random variables on the given probability space, with the inner product between two variables A and B being $E(AB)$.] ◇

2. Properties of orthogonal projections

This section develops properties of orthogonal projections. The results will prove to be useful in analyzing the GLM.

Let $(V, \langle \cdot, \cdot \rangle)$ be an inner product space, let M be a subspace of V, and let

$$M^\perp = \{\, x \in V : x \perp M \,\} \equiv \{\, x \in V : x \perp y \text{ for all } y \in M \,\} \qquad (2.1)$$

be the *orthogonal complement* of M. Note that $M^\perp$ is a subspace of V. Let P_M and Q_M denote orthogonal projection onto M and $M^\perp$, respectively, and let $x \in V$. By Theorem 1.8,

$$x = P_M x + Q_M x$$

is the unique representation of x as the sum of an element of M and an element of $M^\perp$.

2.2 Exercise [2]. Let M and $M_1, \ldots, M_k$ be subspaces of V. Show that

$$d(M^\perp) = d(V) - d(M),$$

$$(M^\perp)^\perp = M,$$

$$M_1 \subset M_2 \iff M_1^\perp \supset M_2^\perp,$$

$$\left(\sum_{1 \le i \le k} M_i\right)^\perp = \bigcap_{1 \le i \le k} M_i^\perp,$$

$$\left(\bigcap_{1 \le i \le k} M_i\right)^\perp = \sum_{1 \le i \le k} M_i^\perp. \qquad \diamond$$

2A. Characterization of orthogonal projections

Let $T \colon V \to V$ be a linear transformation. T is said to be *idempotent* if $T^2 = T$. T is said to be *self-adjoint* if

$$\langle Tx, y \rangle = \langle x, Ty \rangle \qquad (2.3)$$

for all x, y in V.

2.4 Proposition. (i) P_M *is an idempotent, self-adjoint, linear transformation with range* M *and null space* $M^\perp$. (ii) *Conversely, an idempotent, self-adjoint, linear transformation* T *mapping* V *into* V *is the orthogonal projection onto its range.*

Proof. (i) *holds:* Write P for P_M and Q for Q_M. The properties of P_M are all immediate consequences of the orthogonality of M and $M^\perp$ and the uniqueness of the decompositions

$$x = Px + Qx \qquad (Px \in M, \quad Qx \in M^\perp)$$
$$y = Py + Qy \qquad (Py \in M, \quad Qy \in M^\perp),$$

to wit:

(a) P is linear, since $x + y = Px + Qx + Py + Qy$ implies $P(x + y) = Px + Py$ and $cx = cPx + cQx$ implies $P(cx) = c(Px)$;

(b) P is idempotent, since $Px = Px + 0$ implies $P^2 x = Px$;

(c) P is self-adjoint, since $\langle Px, y \rangle = \langle Px, Py + Qy \rangle = \langle Px, Py \rangle = \langle Px + Qx, Py \rangle = \langle x, Py \rangle$ for all $x, y \in V$;

(d) M is the range of P, since on the one hand $x \in V$ implies $Px \in M$ and on the other hand $x \in M$ implies $x = Px$;

(e) $M^{\perp}$ is the null space of P, since $x \in M^{\perp} \Longleftrightarrow x = Qx \Longleftrightarrow Px = 0$.

(ii) *holds*: For $x \in V$ write

$$x = Tx + (x - Tx).$$

Since $Tx \in \mathcal{R}(T)$, we need only show $x - Tx \in \big(\mathcal{R}(T)\big)^{\perp}$. For this observe that for all $y \in V$, one has $\langle x - Tx, Ty \rangle = \langle T(x - Tx), y \rangle = \langle Tx - T^2 x, y \rangle = \langle Tx - Tx, y \rangle = 0$. ∎

2.5 Exercise [2]. Given $0 \neq m \in M$, show that the transformation

$$T: v \to \frac{\langle v, m \rangle}{\langle m, m \rangle} m$$

has all the properties required to characterize it as orthogonal projection onto $[m]$. ◇

2.6 Exercise [2]. Let M be a subspace of V. Show that the orthogonal linear transformation $T: V \to V$ that reflects each $x \in M$ through the origin and leaves each $x \in M^{\perp}$ fixed is $T = I - 2P_M$. ◇

2.7 Exercise [2]. Let $T: V \to V$ be self-adjoint. Show that

$$(\mathcal{R}(T))^{\perp} = \mathcal{N}(T) \tag{2.8}$$

and deduce

$$\mathcal{R}(T^2) = \mathcal{R}(T). \tag{2.9}$$

[Hint: To get (2.9), first show that $\mathcal{N}(T^2) \subset \mathcal{N}(T)$.] ◇

2B. Differences of orthogonal projections

If M and N are two subspaces of V with $M \subset N$, the subspace

$$N - M = \{ x \in N : x \perp M \} = N \cap M^{\perp} \tag{2.10}$$

is called the *(relative) orthogonal complement* of M in N.

2.11 Exercise [2]. Let M and N be subspaces of V with $M \subset N$. Show that if the vectors $v_1, \ldots, v_k$ span N, then the vectors $Q_M v_1, \ldots, Q_M v_k$ span $N - M$. ◇

For self-adjoint transformations S and T mapping V to V, the notation

$$S \leq T \tag{2.12}$$

means $\langle x, Sx \rangle \leq \langle x, Tx \rangle$ for all $x \in V$.

2.13 Proposition. *Let M and N be subspaces of V. The following are equivalent:*

   (i) $P_N - P_M$ *is an orthogonal projection,*

  (ii) $M \subset N$,

 (iii) $P_M \leq P_N$,

 (iv) $\|P_N x\|^2 \geq \|P_M x\|^2$ *for all* $x \in V$,

  (v) $P_M = P_M P_N$,

and

 (vi) $P_N - P_M = P_{N-M}$.

Proof. (ii) *implies* (i) *and* (vi): Suppose $M \subset N$. We will show

$$P_N = P_M + P_{N-M}$$

so that (i) and (vi) hold. $P_M + P_{N-M}$ is a self-adjoint, linear transformation that is idempotent $\left((P_M + P_{N-M})^2 = P_M^2 + P_M P_{N-M} + P_{N-M} P_M + P_{N-M}^2 = P_M + 0 + 0 + P_{N-M}\right)$; its range R is contained in N and yet R contains N because R contains both M and $N - M$. According to Proposition 2.4, these properties characterize $P_M + P_{N-M}$ as P_N.

(i) *implies* (iii): Write P for $P_N - P_M$. For all $x \in V$ one has

$$\langle P_N x, x \rangle - \langle P_M x, x \rangle = \langle (P_N - P_M) x, x \rangle$$
$$= \langle P x, x \rangle = \langle P^2 x, x \rangle = \langle P x, P x \rangle = \|P x\|^2 \geq 0.$$

(iii) *implies* (iv): For all $x \in V$, $\|P_N x\|^2 = \langle P_N x, P_N x \rangle = \langle x, P_N x \rangle \geq \langle x, P_M x \rangle = \|P_M x\|^2$.

(iv) *implies* (v): For all $x \in V$,

$$\|P_M x\|^2 = \|P_N(P_M x)\|^2 + \|Q_N(P_M x)\|^2$$
$$\geq \|P_N(P_M x)\|^2 \geq \|P_M(P_M x)\|^2 = \|P_M x\|^2;$$

this implies $Q_N P_M = 0$. Thus, for all $x, y \in V$,

$$\langle y, P_M Q_N x \rangle = \langle P_M y, Q_N x \rangle = \langle Q_N P_M y, x \rangle = 0,$$

so $P_M Q_N = 0$ also. Now use

$$P_M = P_M(P_N + Q_N) = P_M P_N + P_M Q_N$$

to conclude $P_M = P_M P_N$.

(v) *implies* (ii): Since $P_M = P_M(P_N + Q_N) = P_M P_N + P_M Q_N$, $P_M = P_M P_N$ entails $P_M Q_N = 0 \implies N^\perp \subset M^\perp \implies M \subset N$.

(vi) *implies* (i): Trivial. ∎

2.14 Exercise [3]. Let p and $n_1, \ldots, n_p$ be given and let $V \equiv \{ (x_{ij})_{1 \le j \le n_i, 1 \le i \le p} : x_{ij} \in \mathbb{R} \}$ be endowed with the dot-product

$$\langle (x_{ij}), (y_{ij}) \rangle = \sum_{1 \le i \le p} \sum_{1 \le j \le n_i} x_{ij} y_{ij} .$$

Put

$$M = \{ (x_{ij}) \in V : \text{for some } \beta_1, \ldots, \beta_p \in \mathbb{R}, \; x_{ij} = \beta_i \text{ for all } i, j \} = [v_1, \ldots, v_p]$$

and

$$M_0 = \{ (x_{ij}) \in V : \text{for some } \beta \in \mathbb{R}, \; x_{ij} = \beta \text{ for all } i, j \} = [\sum_{1 \le i \le p} v_i],$$

where $(v_i)_{i'j'} = \delta_{ii'}$. Show that for each $x \in V$,

$$(P_M x)_{ij} = \bar{x}_{i\cdot} \equiv \frac{1}{n_i} \sum_{1 \le j \le n_i} x_{ij} ,$$

$$(P_{M_0} x)_{ij} = \bar{x} \equiv \frac{1}{n} \sum_i n_i \bar{x}_{i\cdot} = \frac{1}{n} \sum_{ij} x_{ij}, \tag{2.15}$$

with $n = \sum_i n_i$, and deduce that

$$(P_{M - M_0} x)_{ij} = \bar{x}_{i\cdot} - \bar{x} . \tag{2.16}$$

[Hint: $v_1, \ldots, v_p$ is an orthogonal basis for M.] $\diamond$

2.17 Exercise [1]. Let M and N be subspaces of an inner product space V. Put $L = M \cap N$. Show that $P_{M-L} P_{N-L} = P_M P_N - P_L$. $\diamond$

2.18 Exercise [2]. Suppose L and M are subspaces of V. Put

$$K = P_L(M) \equiv \{ P_L m : m \in M \}.$$

Show that

$$P_{V-M} - P_{L-K} = P_{(V-M)-(L-K)}. \tag{2.19}$$

Solve this exercise twice: (1) by arguing that $L - K \subset M^\perp$ (in fact, $L - K = L \cap M^\perp$); and (2) by arguing (independently of (1)) that $\|Q_M x\| \ge \|P_{L-K} x\|$ for each $x \in V$. $\diamond$

2.20 Exercise [1]. Suppose $x_1, \ldots, x_\ell$ is a basis for a subspace N of V. Let $M = [x_1, \ldots, x_k]$ be the subspace spanned by the first k of the x_i's, where $1 \le k < \ell$. For an element v of V write

$$P_M v = \sum_{i=1}^k a_i x_i \quad \text{and} \quad P_N v = \sum_{j=1}^\ell b_j x_j, \tag{2.21}$$

and for $j = k+1, \ldots, \ell$ write

$$P_M x_j = \sum_{i=1}^k c_{i,j} x_i; \tag{2.22}$$

the coefficients a_i, b_j, and $c_{i,j}$ in these sums are of course unique. Show that for $i = 1, \ldots, k$

$$a_i = b_i + \sum_{j=k+1}^\ell c_{i,j} b_j. \tag{2.23} \diamond$$

2C. Sums of orthogonal projections

Two subspaces M and N of V are said to be *orthogonal* (or *perpendicular*), written $M \perp N$, if $m \perp n$ for each $m \in M$ and $n \in N$.

2.24 Proposition. *Let* P_1, P_2, ..., P_k *be orthogonal projections onto* M_1, M_2, ..., M_k, *respectively. Set* $P = P_1 + P_2 + \cdots + P_k$. *The following are equivalent:*

(i) P *is an orthogonal projection,*

(ii) $P_i P_j = 0$ *for all* $i \neq j$,

(iii) *the* M_i's *are mutually orthogonal,*

and

(iv) P *is orthogonal projection onto* $\sum_i M_i$.

Proof. (i) *implies* (ii): For each $x \in V$, one has

$$\|x\|^2 \geq \|Px\|^2 = \langle Px, x \rangle = \Big\langle \sum_i P_i x, x \Big\rangle = \sum_i \langle P_i x, x \rangle = \sum_i \|P_i x\|^2.$$

Taking $x = P_j y$ gives

$$\|P_j y\|^2 \geq \|P_j y\|^2 + \sum_{i \neq j} \|P_i P_j y\|^2.$$

As $y \in V$ is arbitrary, we get $P_i P_j = 0$ for $i \neq j$.

(ii) *implies* (iii): $P_i P_j = 0 \implies M_j \subset M_i^\perp \implies M_i \perp M_j$.

(iii) *implies* (i), (iv): For any two orthogonal subspaces K and L of V, part (vi) of Proposition 2.13 gives $P_{K+L} = P_K + P_L$. By induction,

$$P = \sum_i P_i = P_{\sum_i M_i}.$$

(iv) *implies* (i): Trivial.   ■

The most important thing here is that (iii) implies (iv): Given mutually orthogonal subspaces $M_1, \ldots, M_k$,

$$P_{\sum_i M_i} = \sum_i P_{M_i}. \tag{2.25}$$

Notice how this generalizes (1.14).

2.26 Exercise [3]. Let I and J be positive integers and let the space V of $I \times J$ matrices $(x_{ij})_{1 \leq i \leq I, 1 \leq j \leq J}$ be endowed with the dot-product

$$\langle (x_{ij}), (y_{ij}) \rangle = \sum_i \sum_j x_{ij} y_{ij}.$$

Let M_R be the subspace of row-wise constant matrices ($x_{ij} = x_{ij'}$, for all i, j, j'), M_C the subspace of column-wise constant matrices ($x_{ij} = x_{i'j}$, for all i, i', j), and M_G the subspace of constant matrices ($x_{ij} = x_{i'j'}$, for all i, i', j, j'). Show that

$$(P_{M_R} x)_{ij} = \bar{x}_{i\cdot} = \frac{1}{J} \sum_j x_{ij}, \qquad (2.27_R)$$

$$(P_{M_C} x)_{ij} = \bar{x}_{\cdot j} \equiv \frac{1}{I} \sum_i x_{ij}, \qquad (2.27_C)$$

$$(P_{M_G} x)_{ij} = \bar{x}_{\cdot\cdot} \equiv \frac{1}{IJ} \sum_{ij} x_{ij} \qquad (2.27_G)$$

for each $x \in V$. Show further that

$$M_G + (M_R - M_G) + (M_C - M_G)$$

is an orthogonal decomposition of

$$M = M_R + M_C$$

and deduce that

$$(P_M x)_{ij} = \bar{x}_{i\cdot} + \bar{x}_{\cdot j} - \bar{x}_{\cdot\cdot}. \qquad (2.28)$$

[Hint: The $I \times J$ matrices $r_1, \ldots, r_I$ defined by $(r_i)_{i'j'} = \delta_{ii'}$ are an orthogonal basis for M_R.]   ◇

2D. Products of orthogonal projections

Two subspaces M and N of V are said to be *book orthogonal*, written $M \perp_B N$, if $(M - L) \perp (N - L)$ with $L = M \cap N$. The imagery is that of two consecutive pages of a book that has been opened in such a way that the pages are at right angles to one another. Note that $M \perp_B N$ if $M \perp N$, or if $M \subset N$, or if $M \supset N$. In particular, V and the trivial subspace 0 are each book orthogonal to every subspace of V.

2.29 Proposition. *Let M and N be two subspaces of V. The following are equivalent:*

   (i) *$P_M P_N$ is an orthogonal projection,*

   (ii) *P_M and P_N commute,*

   (iii) *$M \perp_B N$,*

and

   (iv) *$P_M P_N = P_L = P_N P_M$ with $L = M \cap N$.*

Proof. (iv) *implies* (i): Trivial.

(i) *implies* (ii): Given that $P_M P_N$ is an orthogonal projection, one has

$$\langle P_M P_N x, y \rangle = \langle x, P_M P_N y \rangle = \langle P_M x, P_N y \rangle = \langle P_N P_M x, y \rangle$$

for all $x, y \in V$, so $P_M P_N = P_N P_M$.

(ii) *implies* (iv): Set $P = P_M P_N$. Using (ii), we have that P is linear, self-adjoint, and idempotent ($P^2 = P_M P_N P_M P_N = P_M^2 P_N^2 = P_M P_N = P$); moreover, $\mathcal{R}(P) = L$, so $P = P_L$ by Proposition 2.4.

(iv) *is equivalent to* (iii): On the one hand, (iv) means $P_M P_N - P_L = 0$ and on the other hand, (iii) is equivalent to $P_{M-L} P_{N-L} = 0$ (use (ii) $\iff$ (iii) in Proposition 2.24). That (iii) and (iv) are equivalent follows from the identity $P_{M-L} P_{N-L} = P_M P_N - P_L$ of Exercise 2.17.   ∎

2.30 Exercise [1]. Show that the subspaces M_R and M_C of Exercise 2.26 are book orthogonal.   ◇

2.31 Exercise [1]. In $V = \mathbb{R}^3$, let $\boldsymbol{x} = (1,0,0)^T$, $\boldsymbol{y} = (0,1,0)^T$, and $\boldsymbol{z} = (0,0,1)^T$. Put $L = [\boldsymbol{x}+\boldsymbol{y}]$, $M = [\boldsymbol{x}, \boldsymbol{y}]$, and $N = [\boldsymbol{y}, \boldsymbol{z}]$. Show that $L \subset M$ and $M \perp_B N$, but it is not the case that $L \perp_B N$.   ◇

2.32 Exercise [2]. Suppose that $L_1, \ldots, L_k$ are mutually orthogonal subspaces of V. Show that the subspaces

$$M_J = \sum\nolimits_{j \in J} L_j \qquad (2.33)$$

of V with $J \subset \{1, 2, \ldots, k\}$ are mutually book orthogonal.   ◇

2.34 Exercise [2]. Show that if M and N are book orthogonal subspaces of V, then so are M and $N^\perp$, and so are $M^\perp$ and $N^\perp$.   ◇

2.35 Exercise [2]. Suppose that $M_1, \ldots, M_k$ are mutually book orthogonal subspaces of V. Show that $\prod_{1 \le j \le k} P_{M_j}$ is orthogonal projection onto $\bigcap_{1 \le j \le k} M_j$ and $\prod_{1 \le j \le k} Q_{M_j}$ is orthogonal projection onto $\left(\sum_{1 \le j \le k} M_j\right)^\perp$.   ◇

2.36 Exercise [2]. Show that if $M_1, \ldots, M_k$ are mutually book orthogonal subspaces of V, then so are the subspaces of V of the form $\bigcap_{j \in J} M_j$, where $J \subset \{1, 2, \ldots, k\}$.   ◇

2.37 Exercise [3]. Prove the following converse to Exercise 2.32: If $M_1, \ldots, M_n$ are mutually book orthogonal subspaces of V, then there exist mutually orthogonal subspaces $L_1, \ldots, L_k$ of V such that each M_m can be written in the form (2.33).
[Hint: Multiply the identity $I_V = \prod_{1 \le \ell \le n}(P_{M_\ell} + Q_{M_\ell})$ by P_{M_m} after expanding out the product.]   ◇

2.38 Exercise [4]. Let M and N be arbitrary subspaces of V and put $L = M \cap N$. Let the successive products P_M, $P_N P_M$, $P_M P_N P_M$, $P_N P_M P_N P_M$, ... of P_M and P_N in alternating order be denoted by $T_1, T_2, T_3, T_4, \ldots$. Show that for each $x \in V$,

$$T_j x \to P_L x \quad \text{as} \quad j \to \infty.$$

[Hint: First observe that $T_j - P_L$ is the j-fold product of P_{M-L} and P_{N-L} in alternating order and then make use of Exercise 1.24.]   ◇

2E. An algebraic form of Cochran's theorem

We begin with an elementary lemma:

2.39 Lemma. *Suppose* $M, M_1, M_2, \ldots, M_k$ *are subspaces of* V *with* $M = M_1 + M_2 + \cdots + M_k$. *The following are equivalent:*

(i) *every vector* $x \in M$ *has a unique representation of the form*

$$x = \sum_{i=1}^{k} x_i$$

with $x_i \in M_i$ *for each* i,

(ii) *for* $1 \le i < k$, M_i *and* $\sum_{j>i} M_j$ *are disjoint,*

(iii) $d(M) = \sum_{i=1}^{k} d(M_i)$ $(d \equiv$ *dimension*).

Proof. (i) *is equivalent to* (ii): Each of (i) and (ii) is easily seen to be equivalent to the following property:

$$\sum_{i=1}^{k} z_i = 0 \text{ with } z_i \in M_i \text{ for each } i \text{ implies } z_i = 0 \text{ for each } i.$$

(ii) *is equivalent to* (iii): The identity

$$d(K + L) = d(K) + d(L) - d(K \cap L),$$

holding for arbitrary subspaces K and L of V, gives

$$d\left(M_i + \left(\sum_{j>i} M_j\right)\right) = d(M_i) + d\left(\sum_{j>i} M_j\right) - d\left(M_i \cap \left(\sum_{j>i} M_j\right)\right)$$

for $1 \le i < k$. Combining these relations, we find

$$d(M) = d\left(\sum_{j=1}^{k} M_j\right) = \sum_{j=1}^{k} d(M_j) - \sum_{i=1}^{k-1} d\left(M_i \cap \left(\sum_{j>i} M_j\right)\right).$$

So $d(M) = \sum_{i=1}^{k} d(M_i)$ if and only if $d\left(M_i \cap \left(\sum_{j>i} M_j\right)\right) = 0$ for $1 \le i < k$. ∎

When the equivalent conditions of the lemma are met, one says that M is the *direct sum* of the M_i's, written $M = \oplus_{i=1}^{k} M_i$. The sum M of mutually orthogonal subspaces M_i is necessarily direct, and in this situation one says that the M_i's form an *orthogonal decomposition* of M.

We are now ready for the following result, which forms the algebraic basis for Cochran's theorem (see Section 3.9) on the distribution of quadratic forms in normally distributed random variables. There is evidently considerable overlap between the result here and Proposition 2.24. Note though that here it is the *sum* of the operators involved that is assumed at the outset to be an orthogonal projection, whereas in Proposition 2.24 it is the individual *summands* that are postulated to be orthogonal projections.

2.40 Proposition (Algebraic form of Cochran's theorem). Let $T_1, \ldots, T_k$ be *self-adjoint, linear transformations of V into V and suppose that*

$$P = \sum_{1 \leq i \leq k} T_i$$

is an orthogonal projection. Put $M_i = \mathcal{R}(T_i)$, $1 \leq i \leq k$, and $M = \mathcal{R}(P)$. The following are equivalent:

> (i) *each T_i is idempotent,*
>
> (ii) $T_i T_j = 0$ *for $i \neq j$,*
>
> (iii) $\sum_{i=1}^{k} d(M_i) = d(M)$,

and

> (iv) $M_1, \ldots, M_k$ *form an orthogonal decomposition of M.*

Proof. We treat the case where P is the identity transformation I on V, leaving the general case to the reader as an exercise.

(i) *implies* (iv): Each T_i is an orthogonal projection by (i) and $\sum_i T_i$ an orthogonal projection by the umbrella assumption $\sum_i T_i = I$. Proposition 2.24 says that $\sum_i T_i$ is orthogonal projection onto the orthogonal direct sum $\oplus_i M_i$. But $\sum_i T_i = I$ is trivially orthogonal projection onto V. Thus the M_i's form an orthogonal decomposition of V.

(iv) *implies* (iii): Trivial.

(iii) *implies* (ii): For any $x \in V$,

$$x = Ix = \sum_i T_i x,$$

so $V = \sum_{i=1}^{k} M_i$. Taking $x = T_j y$ and using Lemma 2.39 with $M = V$, we get

$$T_i T_j y = 0 \quad \text{for } i = 1, \ldots, j-1, j+1, \ldots, k$$

(as well as $T_j y = T_j^2 y$ — but we do not need this now). Since j and y are arbitrary, $T_i T_j = 0$ for $i \neq j$.

(ii) *implies* (i): One has, using (ii),

$$T_i - T_i^2 = T_i(I - T_i) = T_i\left(\sum_{j \neq i} T_j\right) = \sum_{j \neq i} T_i T_j = 0$$

for all i.   ∎

2.41 Exercise [3]. Prove Proposition 2.40 for the general orthogonal projection P. [Hint: The case of general P can be reduced to the case $P = I$, which was treated in the preceding proof, by introducing $T_0 = Q_M$. Specifically, let the conditions (i'), ..., (iv') be defined like (i), ..., (iv), but with the indices ranging from 0 to k instead of from 1 to k, and with M replaced by V. The idea is to show that condition (c') is equivalent to condition (c), for $c = $ i, ..., iv. In arguing that (ii) implies (ii'), make use of Exercise 2.7.]◇

2.42 Exercise [3]. In the context of $\mathbb{R}^n$, suppose $I = T_1 + T_2$ with T_1 being the transformation having the matrix

$$(t_{ij}) = \begin{pmatrix} 1/n & 1/n & \ldots & 1/n \\ 1/n & 1/n & \ldots & 1/n \\ \vdots & \vdots & & \vdots \\ 1/n & 1/n & \ldots & 1/n \end{pmatrix}$$

with respect to the usual coordinate basis, so that

$$[T_1((x_j))]_i = \sum_{1 \leq j \leq n} t_{ij} x_j$$

for $1 \leq i \leq n$. In regard to Proposition 2.40, show that T_1 and T_2 are self-adjoint and that some one (and therefore all) of conditions (i)–(iv) is satisfied.   ◇

2.43 Exercise [2]. In the context of $\mathbb{R}^2$, let T_1 and T_2 be the transformations having the matrices

$$[T_1] = \begin{pmatrix} 0 & 1 \\ 1 & 0 \end{pmatrix} \quad \text{and} \quad [T_2] = \begin{pmatrix} 1 & -1 \\ -1 & 1 \end{pmatrix}$$

with respect to the usual coordinate basis. Show that even though T_1 and T_2 are self-adjoint and sum to the identity, some one (and therefore each) of conditions (i)–(iv) of Proposition 2.40 is not satisfied.   ◇

3. Tjur's theorem

Let V be an inner product space. A finite collection $\mathcal{L}$ of distinct subspaces of V such that

(T1) $L_1, L_2 \in \mathcal{L}$ implies L_1 and L_2 are book orthogonal,

(T2) $L_1, L_2 \in \mathcal{L}$ implies $L_1 \cap L_2 \in \mathcal{L}$, and   (3.1)

(T3) $V \in \mathcal{L}$

is called a *Tjur system* (after the Danish statistician Tue Tjur). For example, the subspaces V, M, and M_0 of Exercise 2.14 constitute a Tjur system, as do the subspaces V, M_R, M_C, and M_G of Exercise 2.26. Since intersections of mutually book orthogonal spaces are themselves mutually book orthogonal (see Exercise 2.36), any finite collection of mutually book orthogonal subspaces of V can be augmented to a Tjur system.

The elements of a Tjur system $\mathcal{L}$ are partially ordered by inclusion. For $L \in \mathcal{L}$ we write $K \leq L$ to mean that $K \in \mathcal{L}$ and K is a subset of L, and we write $K < L$ to mean that $K \leq L$ and $K \neq L$. This notation is used in the following key theorem, which shows that the elements of $\mathcal{L}$ can be represented neatly and simply in terms of an explicit orthogonal decomposition of V.

3.2 Theorem (Tjur's theorem). *Let $\mathcal{L}$ be a Tjur system of subspaces of V. For each $L \in \mathcal{L}$, put*

$$L^\circ = L - \sum_{K < L} K. \tag{3.3}$$

Then

(i) *the subspaces L° for $L \in \mathcal{L}$ are mutually orthogonal,*

(ii) $V = \sum_{L \in \mathcal{L}} L^\circ$,

(iii) *for each $L \in \mathcal{L}$, one has $L = \sum_{K \le L} K^\circ$.*

Moreover, the L°'s are uniquely determined by these conditions.

Proof. (i) *and* (iii) *imply* (3.3): Suppose that $\{ L^* : L \in \mathcal{L} \}$ is a collection of mutually orthogonal subspaces of V such that $L = \sum_{K \le L} K^*$ for each $L \in \mathcal{L}$; we claim $L^* = L^\circ$ for each L. For this note that $L^* + \left(\sum_{K < L} K^* \right)$ is an orthogonal decomposition of L, so

$$L^* = L - \left(\sum_{K < L} K^* \right) = L - \left(\sum_{K < L} K \right) = L^\circ.$$

(3.3) *implies* (i)–(iii): We first use induction to show that

$$L = \sum_{K \le L} K^\circ \tag{3.4}$$

for each $L \in \mathcal{L}$. Condition (T2) implies that $\mathcal{L}$ has a smallest element, say L_0; (3.4) holds for L_0 because $L_0^\circ = L_0$. Suppose now

$$K = \sum_{J \le K} J^\circ \quad \text{for all } K \in \mathcal{L} \text{ with } K < L. \tag{3.5}$$

Then

$$L = L^\circ + \sum_{K < L} K \qquad \text{(by the definition (3.3) of } L^\circ)$$
$$= L^\circ + \sum_{K < L} \left(\sum_{J \le K} J^\circ \right) \qquad \text{(by the induction hypothesis (3.5))}$$
$$= L^\circ + \sum_{J < L} J^\circ = \sum_{K \le L} K^\circ,$$

so (3.4) holds for L. Claim (iii) now follows from Exercise 3.7 below, and (ii) holds because $V \in \mathcal{L}$ by (T3).

To complete the proof we show that (i) holds. Let $L_1, L_2 \in \mathcal{L}$; we have to show that

$$L_1^\circ \perp L_2^\circ. \tag{3.6}$$

By (T2), $L = L_1 \cap L_2 \in \mathcal{L}$. There are three cases to consider: (1) $L < L_1$ and $L < L_2$; (2) $L = L_1$; and (3) $L = L_2$. In case (1), $L_1^\circ \subset L_1 - L$ and $L_2^\circ \subset L_2 - L$, so (3.6) holds because L_1 and L_2 are book orthogonal by (T1). In case (2), $L_1 < L_2$, so (3.6) holds because $L_1^\circ \subset L_1$ and $L_2^\circ \subset L_2 - L_1$. Case (3) is like case (2). ∎

3.7 Exercise *(The principle of induction for a partially ordered set)* [4]. Let I be a finite set and let $\leq$ be a *partial order* on I:

$$i \leq i \quad \text{for all } i \in I, \tag{3.8_1}$$

$$i \leq j \text{ and } j \leq k \quad \text{implies} \quad i \leq k, \tag{3.8_2}$$

$$i \leq j \text{ and } j \leq i \quad \text{implies} \quad i = j. \tag{3.8_3}$$

For $i, j \in I$, write $i < j$ to mean $i \leq j$ and $i \neq j$. Say that an element j of I is *minimal* if there does not exist an $i \in I$ such that $i < j$. For each $i \in I$, let $S(i)$ be a statement involving i. Prove the *principle of induction for I*: If $S(i)$ is valid for each minimal element i, and if for each nonminimal element j, $S(j)$ is valid whenever $S(i)$ is valid for all $i < j$, then $S(j)$ is valid for all $j \in I$.
[Hint: If $i < j$ and $j < k$, then i, j, and k must be distinct elements of I.]   ◇

3.9 Exercise [2]. Show that the subspace L° defined by (3.3) can be written as $L \cap \left(\bigcap_{K < L}(L - K) \right)$.
[Hint: See Exercise 2.2.]   ◇

3.10 Exercise [2]. Show by example that the conclusion (i) of Tjur's theorem would not follow if condition (T2) were dropped from (3.1).   ◇

3.11 Exercise [1]. Use Tjur's Theorem to solve Exercise 2.37.   ◇

Tjur's theorem states in part that $\sum_{L \in \mathcal{L}} L^\circ$ is an orthogonal decomposition of V. In applications one needs to know how to project onto the subspaces L°, what dimensions these subspaces have, and how long the components $P_{L^\circ} v$ of a vector $v \in V$ are. This information is readily obtained from corresponding information about the original spaces $L \in \mathcal{L}$. Indeed, conclusion (iii) in Tjur's theorem implies that

$$P_L = \sum_{K \leq L} P_{K^\circ} \quad \text{for } L \in \mathcal{L}. \tag{3.12}$$

These equations can be solved recursively for the P_{L°'s, working upward from the minimal element L_0 of $\mathcal{L}$ to the maximal element V:

$$\begin{aligned} P_{L_0^\circ} &= P_{L_0}, \\ P_{L^\circ} &= P_L - \sum_{K < L} P_{K^\circ} \quad \text{for } L > L_0. \end{aligned} \tag{3.13}$$

Similarly, the dimensions of the L° spaces can be found by solving the equations

$$d(L) = \sum_{K \leq L} d(K^\circ), \quad L \in \mathcal{L}, \tag{3.14}$$

and for $v \in V$ the squared lengths

$$\ell_{L^\circ}^2 \equiv \ell_{L^\circ}^2 v = \|P_{L^\circ} v\|^2 \tag{3.15}$$

can be found from the corresponding squared lengths

$$\ell_L^2 \equiv \ell_L^2(v) = \|P_L v\|^2 \tag{3.16}$$

by solving the equations

$$\ell_L^2 = \sum_{K \leq L} \ell_{K^\circ}^2, \quad L \in \mathcal{L}. \tag{3.17}$$

In specific situations the computations are facilitated by referring to a *structure diagram*, which displays the ordering of the elements of $\mathcal{L}$. The following example illustrates the method.

3.18 Example. Consider the Tjur system $\{V, M_R, M_C, M_G\}$ of Exercise 2.26. For notational simplicity write R for M_R, C for M_C, and G for M_G. The structure diagram is

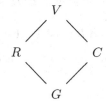

A line between two elements of $\mathcal{L}$ indicates that the lower subspace is included in the upper one. No line needs to be drawn for the inclusion $G \subset V$, since that relation is readily inferred from the diagram as it stands.

Formulas for the projections P_L were given in Exercise 2.26 (see (2.27)). To calculate the P_{L°'s from the P_L's using (3.13), begin by superscripting each L in the structure diagram by P_L. Then, working upward from the bottom of the diagram, subscript each L by P_{L°, calculated as the difference between the corresponding superscript and the sum of all subscripts for spaces K strictly below L (that is, for all K such that $K < L$). For example, the subscript on V is calculated as $P_V - (P_{R^\circ} + P_{C^\circ} + P_{G^\circ})$. The result is

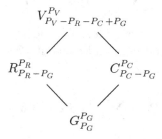

The method of su$^\mathrm{per}_\mathrm{b}$scripts can also be used to organize the computation of the $d(L^\circ)$'s and $\ell_{L^\circ}^2$'s. For example, since $d(V) = IJ$, $d(R) = I$, $d(C) = J$, and $d(G) = 1$, the annotated structure diagram for the calculation of the $d(L^\circ)$'s is

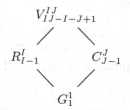

The *analysis of variance table* for a Tjur system $\mathcal{L}$ lists for each $L \in \mathcal{L}$ the quantities L°, $d(L^\circ)$, and $\ell_{L^\circ}^2 = \|P_{L^\circ}v\|^2$ appearing in the decompositions

$$V = \sum\nolimits_{L \in \mathcal{L}} L^\circ, \quad d(V) = \sum\nolimits_{L \in \mathcal{L}} d(L^\circ), \quad \|v\|^2 = \sum\nolimits_{L \in \mathcal{L}} \ell_{L^\circ}^2(v).$$

Following conventional statistical practice, rows for smaller L's (used in building simpler models) appear above rows for larger L's (used in building more complicated models). The analysis of variance table for the Tjur system in the preceding Example is given in Table 3.19.

3.19 Table. *Analysis of variance table for the Tjur system* $\{V, R, C, G\}$ *of Example* 3.18. *The quantities* $\ell_L^2 = \|P_L v\|^2$ *appearing in the "squared length" column are given by*

$$\ell_V^2 = \sum\nolimits_{ij} v_{ij}^2, \quad \ell_R^2 = J \sum\nolimits_i \bar{v}_{i\cdot}^2, \quad \ell_C^2 = I \sum\nolimits_j \bar{v}_{\cdot j}^2, \quad \ell_G^2 = IJ\bar{v}_{\cdot\cdot}^2.$$

Label	Subspace	Dimension	Squared length
G	$G^\circ = G$	1	$\ell_G^2 = IJ\bar{v}_{\cdot\cdot}^2$
R	$R^\circ = R - G$	$I - 1$	$\ell_R^2 - \ell_G^2 = J \sum_i (\bar{v}_{i\cdot} - \bar{v}_{\cdot\cdot})^2$
C	$C^\circ = C - G$	$J - 1$	$\ell_C^2 - \ell_G^2 = I \sum_j (\bar{v}_{\cdot j} - \bar{v}_{\cdot\cdot})^2$
V	$V^\circ = V - (R + C + G)$	$IJ - I - J + 1$	$\ell_V^2 - \ell_R^2 - \ell_C^2 + \ell_G^2$ $= \sum_{ij} (v_{ij} - \bar{v}_{i\cdot} - \bar{v}_{\cdot j} + \bar{v}_{\cdot\cdot})^2$
Sum	V	IJ	$\ell_V^2 = \sum_{ij} v_{ij}^2$

3.20 Exercise [2]. Verify the entries in the "squared lengths" column of Table 3.19. To get the expressions to the left of the $=$ signs, use the method of $\text{su}_{\text{b}}^{\text{per}}$scripts. To get the expressions to the right of the $=$ signs, use the formulas for the P_{L°'s derived in Example 3.18.   ◇

3.21 Exercise [2]. Draw the structure diagram for the Tjur system $\{V, M, M_0\}$ of Exercise 2.14, and use the method of $\text{su}_{\text{b}}^{\text{per}}$scripts to construct the corresponding analysis of variance table.   ◇

3.22 Exercise *(The Möbius inversion formula)* [4]. As in Exercise 3.7, let $\leq$ be a partial order on a finite set I. (1) The *Möbius function* of $(I, \leq)$ is the matrix $\mu = (\mu_{ik})_{i \in I, k \in I}$ such that $\mu_{ik} = 0$ for $i \not\leq k$ and such that

$$\sum\nolimits_{j\,:\,i \leq j \leq k} \mu_{ij} = \delta_{ik} \quad \text{for } i \leq k \tag{3.23}$$

or, equivalently,

$$\sum\nolimits_{j\,:\,i \leq j \leq k} \mu_{jk} = \delta_{ik} \quad \text{for } i \leq k. \tag{3.24}$$

Prove that such a matrix μ exists and is inverse to the matrix $\zeta = (\zeta_{ik})_{i \in I, k \in I}$ with

$$\zeta_{ik} = \begin{cases} 1, & \text{if } i \le k, \\ 0, & \text{otherwise;} \end{cases}$$

ζ is called the *zeta function* of $(I, \le)$. (2) Prove of functions f and g mapping I into some vector space W that

$$g(j) = \sum_{i \le j} f(i) \quad \text{for all } j \in I \tag{3.25}$$

if and only if

$$f(j) = \sum_{i \le j} g(i) \mu_{ij} \quad \text{for } j \in I; \tag{3.26}$$

this result is called the *Möbius inversion formula*.

[Hint: For each fixed $i \in I$, equation (3.23) can be solved recursively for μ_{ik} with $i \le k$.] $\diamond$

3.27 Exercise [3]. Recall that the elements of a Tjur system $\mathcal{L}$ of subspaces of V are partially ordered by inclusion ($\le$). Let μ be the Möbius function for $(\mathcal{L}, \le)$. Use the Möbius inversion formula from the preceding exercise to show that

$$P_{L^\circ} = \sum P_K \mu_{KL}, \quad d(L^\circ) = \sum d(K) \mu_{KL}, \quad \ell_{L^\circ}^2 = \sum \ell_K^2 \mu_{KL} \tag{3.28}$$

for all $L \in \mathcal{L}$, the summations in each case extending over $K \le L$ or even over all $K \in \mathcal{L}$. Check (3.28) against the entries in Table 3.19. $\diamond$

As Example 3.18 suggests, Tjur systems arise naturally in the theory of the design of experiments. The rest of this section elaborates on this idea. In particular, we will see how the notions of book orthogonality, intersection, and inclusion of subspaces correspond to certain simple relationships among the levels of the treatments assigned to the experimental units.

The design of an experiment often calls for the available experimental units (for example, plots of land) to be divided into groups, with each unit in a group receiving the same level of some experimental treatment (for example, type of seed). To abstract this idea, let X be a finite set whose elements are called *experimental units*. A (treatment) *factor* F is a partition of X into nonempty disjoint subsets called *blocks*, or *levels*, of F. The factor U whose blocks are single units is called the *units factor*, while the factor T having only one block is called the *trivial factor*.

For a subset f of X put

$$|f| = \text{cardinality}(f), \tag{3.29}$$

the number of experimental units in f. A factor F is said to be *balanced* if each of its levels is assigned the same number of units, that is, if

$$|f| = |g| \quad \text{for all blocks } f \text{ and } g \text{ of } F. \tag{3.30}$$

For example, if $X = \{ (i, j) : 1 \le i \le I, \ 1 \le j \le J \}$, then the row factor R with blocks $r_i = \{ (i, j) : 1 \le j \le J \}$ for $1 \le i \le I$ is balanced because $|r_i| = J$ for each i.

If F is a factor and x and y are experimental units, the notation

$$x \sim_F y \qquad (3.31)$$

means that x and y are assigned the same level of F, that is, belong to the same block of F. The relation $\sim_F$ is an equivalence relation:

$$x \sim_F x \quad \text{for each } x \in X, \qquad (3.32_1)$$

$$x \sim_F y \quad \text{implies} \quad y \sim_F x, \qquad (3.32_2)$$

$$x \sim_F y \text{ and } y \sim_F z \quad \text{implies} \quad x \sim_F z. \qquad (3.32_3)$$

Conversely, if $\sim$ is an arbitrary equivalence relation on X, then its equivalence classes constitute a factor. The correspondence between factors and equivalence relations is one-to-one.

Let F and G be factors. One says F *is nested in* G if each block of G is a union of blocks of F or, equivalently, if

$$\text{for } x, y \in X, \ x \sim_F y \text{ implies } x \sim_G y; \qquad (3.33)$$

this situation is written as

$$G \leq F \qquad (3.34)$$

because it implies that F has at least as many levels as G. The units factor U is nested in each factor, and each factor is nested in the trivial factor T. The relation $\leq$ is a partial order on the set of factors of X:

$$F \leq F \quad \text{for all factors } F, \qquad (3.35_1)$$

$$H \leq G \text{ and } G \leq F \quad \text{implies} \quad H \leq F, \qquad (3.35_2)$$

$$G \leq F \text{ and } F \leq G \quad \text{implies} \quad F = G. \qquad (3.35_3)$$

Interesting factors can be built up from given ones by means of their equivalence relations. For example, suppose F and G are factors. Write

$$x \sim_\times y \qquad (3.36)$$

for $x, y \in X$ to mean $x \sim_F y$ and $x \sim_G y$. $\sim_\times$ is an equivalence relation; the factor $F \times G$ whose blocks are the $\sim_\times$-equivalence classes is called the *cross-classification induced by F and G*. It turns out that $F \times G$ is the least upper bound of F and G with respect to the "nested in" ordering, so $F \times G$ is also called the *maximum of F and G* and written as $F \vee G$.

3.37 Exercise [2]. Let F and G be factors. Show that $F \times G$ is indeed the *least upper bound* of F and G with respect to the ordering (3.34):

$$F \leq F \times G \quad \text{and} \quad G \leq F \times G, \qquad (3.38_1)$$

$$F \leq H \text{ and } G \leq H \quad \text{implies} \quad F \times G \leq H. \qquad (3.38_2) \ \diamond$$

Let again F and G be factors. Write

$$x \sim_\wedge y \tag{3.39}$$

for $x, y \in X$ to mean that there exists a finite sequence $z_0, \ldots, z_k$ of experimental units such that

$$z_0 = x \quad \text{and} \quad z_k = y \tag{3.40_1}$$

and

$$z_{j-1} \sim_F z_j \quad \text{and/or} \quad z_{j-1} \sim_G z_j \quad \text{for } 1 \le j \le k. \tag{3.40_2}$$

Loosely speaking, $x \sim_\wedge y$ if it is possible to move from x to y through X in a finite number of steps, each of which takes place within a block of F (an F-step) or a block of G (a G-step); an efficient move would alternate F-steps with G-steps, but this is not required. $\sim_\wedge$ is an equivalence relation; the factor whose blocks are the $\sim_\wedge$-equivalence classes turns out to be the greatest lower bound of F and G, and so it is called the *minimum of F and G* and written as $F \wedge G$.

3.41 Exercise [3]. Let F and G be factors. Verify that the relation $\sim_\wedge$ is indeed an equivalence relation and that the factor $F \wedge G$ is the *greatest lower bound* of F and G:

$$F \wedge G \le F \quad \text{and} \quad F \wedge G \le G, \tag{3.42_1}$$

$$H \le F \text{ and } H \le G \quad \text{implies} \quad H \le F \wedge G. \tag{3.42_2} \diamond$$

3.43 Example. Suppose $X = \{1, 2, 3, 4, 5, 6, 7, 8, 9\}$ and the experimental units are allocated to the 3 levels r_1, r_2, and r_3 of a row factor R and the 3 levels c_1, c_2, and c_3 of a column factor C as indicated below.

Rows	Columns		
	c_1	c_2	c_3
r_1	1,2	3	
r_2	4,5	6	9
r_3			7,8

For example, units 1, 2, and 3 are assigned to level r_1 of R, and units 3 and 6 are assigned to level c_2 of C. The blocks of $R \times C$ are $\{1, 2\}$, $\{3\}$, $\{4, 5\}$, $\{6\}$, $\{7, 8\}$, and $\{9\}$. $R \wedge C$ is the trivial factor because each unit can be reached from each other unit by a succession of row and column steps. If the last experimental unit 9 were to be removed from X, then $R \wedge C$ would have 2 blocks, $\{1, 2, 3, 4, 5, 6\}$ and $\{7, 8\}$. •

3.44 Exercise [4]. Let $F_1, \ldots, F_k$ be factors such that $f_1 \cap \cdots \cap f_k \neq \emptyset$ for all $f_1 \in F_1$, $\ldots$, $f_k \in F_k$. For subsets J of $K \equiv \{1, \ldots, k\}$, let

$$F_J = \prod_{j \in J} F_j \tag{3.45}$$

denote the cross-classification induced by factors F_j with $j \in J$ — $x \sim_{F_J} y$ if and only if $x \sim_{F_j} y$ for all $j \in J$. (Take $F_{\emptyset}$ to be trivial factor T.) Show that

$$F_{J_1} \wedge F_{J_2} = F_{J_1 \cap J_2} \tag{3.46}$$

for all subsets J_1 and J_2 of K.

[Hint First do the case where $k = 3$, $J_1 = \{1, 2\}$, and $J_2 = \{2, 3\}$.]   ◇

Now let V be the vector space of real-valued functions on X. For each subset f of X, let $I_f \in V$ be the *indicator function of f*:

$$I_f(x) = \begin{cases} 1, & \text{if } x \in f, \\ 0, & \text{if } x \notin f. \end{cases} \tag{3.47}$$

For each factor F, let

$$L_F = [\, I_f : f \in F \,] = \{\, v \in V : x \sim_F y \text{ implies } v(x) = v(y) \,\} \tag{3.48}$$

be the subspace of V consisting of functions that are constant on each block of F. Note that L_F has dimension

$$d(L_F) = |F|, \quad \text{the number of blocks of } F. \tag{3.49}$$

3.50 Proposition. *Let F and G be factors. Then*

$$G \leq F \quad \text{if and only if} \quad L_G \subset L_F \tag{3.51}$$

and

$$L_{F \wedge G} = L_F \cap L_G. \tag{3.52}$$

Proof. *$G \leq F$ implies $L_G \subset L_F$:* $G \leq F$ means that each block of G is a union of blocks of F, so constancy on blocks of G trivially implies constancy on blocks of F.

$L_G \subset L_F$ implies $G \leq F$: Suppose $L_G \subset L_F$. Let g be a block of G and f be a block of F such that $f \cap g \neq \emptyset$. We have to show that $f \subset g$. For this, note that the function $I_g \in L_G \subset L_F$ has the value 1 at some point of f and so is identically 1 on f.

$L_{F \wedge G} \subset L_F \cap L_G$: This follows from (3.51) applied to $F \wedge G \leq F$ and $F \wedge G \leq G$.

$L_F \cap L_G \subset L_{F \wedge G}$: Suppose $v \in V$ is constant over blocks of both F and G. We have to show v is constant over blocks of $F \wedge G$. For this, suppose $x, y \in X$ with $x \sim_{F \wedge G} y$. Choose $z_0, \ldots, z_k$ satisfying (3.40). Then $v(z_{j-1}) = v(z_j)$ for each j, so $v(x) = v(z_0) = v(z_k) = v(y)$.   ∎

Now let V be endowed with the dot-product

$$\langle u, v \rangle = \sum_{x \in X} u(x) v(x). \tag{3.53}$$

Let P_F denote orthogonal projection onto L_F. Since the functions I_f for $f \in F$ are an orthogonal basis for L_F, we have

$$P_F v = \sum_{f \in F} \bar{v}_f I_f, \qquad (3.54)$$

where

$$\bar{v}_f = \frac{\langle I_f, v \rangle}{\langle I_f, I_f \rangle} = \frac{1}{|f|} \sum_{x \in F} v(x); \qquad (3.55)$$

in other words, the value of $P_F v$ at an experimental unit y is the average value of $v \in V$ over all units x that are in the same block of F as y. In situations where the symbol v is adorned with subscripts and/or superscripts, we will sometimes write $A_f(v)$ in place of $\bar{v}_f$. Note that

$$\ell^2_{L_F}(v) \equiv \|P_F v\|^2 = \sum_{x \in X} \left((P_F v)(x) \right)^2 = \sum_{f \in F} |f| \bar{v}_f^2 \equiv SS_F, \qquad (3.56)$$

where the symbol SS denotes *sum of squares*.

Two factors F and G are said be *orthogonal* if L_F and L_G are book orthogonal. By Proposition 2.29, this is equivalent to $P_F P_G = P_G P_F$, and also to $P_F P_G = P_{L_F \cap L_G}$. In view of (3.52), this last condition is the same as $P_F P_G = P_{F \wedge G}$. The following proposition gives a useful characterization of orthogonality in terms of the sizes of the blocks of F, G, and $F \wedge G$. If f and h are subsets of X, say that f *is nested in* h if f is a subset of h.

3.57 Proposition. *The factors F and G are orthogonal if and only if*

$$\frac{|f \cap g|}{|h|} = \frac{|f|}{|h|} \times \frac{|g|}{|h|} \qquad (3.58)$$

for all blocks $f \in F$, $g \in G$, and $h \in H \equiv F \wedge G$ such that f and g are nested in h.

Proof. The functions $\delta_x = I_{\{x\}}$, $x \in X$, form a basis for V, so F and G are orthogonal if and only if $\langle \delta_x, P_H \delta_y \rangle = \langle \delta_x, P_F P_G \delta_y \rangle$, and hence if and only if

$$\langle \delta_x, P_H \delta_y \rangle \equiv \langle P_F \delta_x, P_G \delta_y \rangle \qquad (3.59)$$

for all $x, y \in X$. The left- and right-hand sides of (3.59) are 0 unless x and y both belong to the same block h of H, in which case the left-hand side is

$$\left\langle \delta_x, \sum_{\tilde{h} \in H} A_{\tilde{h}}(\delta_y) I_{\tilde{h}} \right\rangle = A_h(\delta_y) = 1/|h|$$

and the right-hand is

$$\left\langle \sum_{\tilde{f} \in F} A_{\tilde{f}}(\delta_x) I_{\tilde{f}}, \sum_{\tilde{g} \in G} A_{\tilde{g}}(\delta_y) I_{\tilde{g}} \right\rangle = A_f(\delta_x) A_g(\delta_y) \langle I_f, I_g \rangle = \frac{|f \cap g|}{|f| |g|},$$

where f is the block of F containing x and g is the block of G containing y; necessarily, $f \subset h$ and $g \subset h$. ∎

3.60 Exercise [2]. Show of factors F and G that $G \leq F$ implies that F and G are orthogonal. Solve this exercise twice, once using (3.51) and once using (3.58). ◇

The following exercise gives a convenient reformulation of the orthogonality criterion (3.58).

3.61 Exercise [3]. Let F and G be factors. Show that F and G are orthogonal if and only if:

For each level h of $H \equiv F \wedge G$ and levels g_1 and g_2 of G nested in h, there exists a constant $c = c(h, g_1, g_2)$ such that $|f \cap g_1| = c|f \cap g_2|$ for all levels f (3.62) of F nested in h,

this being the so-called *condition of proportional cell counts*. ◇

3.63 Exercise [2]. Each of the four tables below gives the number of experimental units to be allocated to the cells of a 3×3 rectangular array. In which cases is the implied row factor R orthogonal to the implied column factor C?

Table 1			Table 2			Table 3			Table 4		
1	1	1	3	3	3	2	1	0	2	1	0
1	1	1	2	2	2	2	1	1	2	1	0
1	1	1	1	1	1	0	0	2	0	0	2

◇

3.64 Exercise [2]. Consider a $2 \times 2 \times 2$ design having a row factor R with levels r_1 and r_2, a column factor C with levels c_1 and c_2, and a height factor H with levels h_1 and h_2. Suppose the 10 experimental units are allocated in such a way that

$$|r_i \cap c_j \cap h_k| = \begin{cases} 2, & \text{if } (i,j,k) = (1,1,2) \text{ or } (2,2,2), \\ 1, & \text{otherwise.} \end{cases}$$

Show that $R \perp H$ and $C \perp H$, but $(R \times C) \not\perp H$. Is $(R \wedge C) \perp H$? ◇

A *Tjur design* $\mathcal{D}$ is a collection of distinct factors such that

(D1) $F, G \in \mathcal{D}$ implies F and G are orthogonal,

(D2) $F, G \in \mathcal{D}$ implies $F \wedge G \in \mathcal{D}$, and   (3.65)

(D3) the units factor U is in $\mathcal{D}$.

The subspaces L_F for F in a Tjur design $\mathcal{D}$ constitute a Tjur system $\mathcal{L}$; indeed, (D1) trivially implies (T1), (D2) implies (T2) by (3.52), and (D3) implies (T3) since $L_U = V$. Moreover, by (3.51), the ordering of the factors $F \in \mathcal{D}$ corresponds to the ordering of the L_F's. The quantities $P_{L_F} \equiv P_F$, $d(L_F)$, and $SS_F = \|P_F v\|^2$ are easily obtained from (3.54), (3.49), and (3.56), respectively, and the analysis of variance table can be written down more or less at sight using a factor structure diagram and the method of $\mathrm{su}_{\mathrm{b}}^{\mathrm{per}}$scripts.

3.66 Exercise *(A split-plot design)* [4]. Suppose that 5 treatments $a_1, \ldots, a_5$ are applied to 15 plots, each treatment being applied to 3 plots. Each plot is subdivided into 2 subplots to which 2 further treatments b_1, b_2 are applied, 1 treatment to each subplot. The relevant factors are: the units factor U (for subplots) with 30 levels; a "plot" factor P with 15 levels, an "a-treatment" factor A with 5 levels, a "b-treatment" factor B with 2 levels, a "combined treatment" factor $A \times B$ with 10 levels, and the trivial factor T with 1 level. Using the following template of •'s to represent the 30 subplots, show one way of assigning the experimental units to these factors:

$$
\begin{array}{ccccccccccc}
\bullet & \bullet & \bullet & \quad & \bullet & \bullet & \bullet & \quad & \bullet & \bullet & \bullet & \quad & \bullet & \bullet & \bullet & \quad & \bullet & \bullet & \bullet \\
\bullet & \bullet & \bullet & \quad & \bullet & \bullet & \bullet & \quad & \bullet & \bullet & \bullet & \quad & \bullet & \bullet & \bullet & \quad & \bullet & \bullet & \bullet
\end{array}
$$

Verify that $\{U, P, A, B, A \times B, T\}$ is a Tjur design with the factor structure diagram

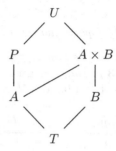

and use the method of $\mathrm{su}_{\mathrm{b}}^{\mathrm{per}}$scripts to compute the corresponding analysis of variance table.   ◇

4. Self-adjoint transformations and the spectral theorem

Recall that a linear transformation $T: V \rightarrow V$ such that

$$
\langle Tx, y \rangle = \langle x, Ty \rangle \quad \text{for all } x, y \in V \tag{4.1}
$$

is said to be *self-adjoint*. In this section we are going to study the geometry of self-adjoint transformations.

A few words are in order about one reason why self-adjoint transformations are of special interest in the study of the GLM. Consider the canonical example $(V, \langle \cdot, \cdot \rangle) = (\mathbb{R}^n, \text{dot product})$. Suppose $\boldsymbol{Y}^{n \times 1} = (Y_i)_{1 \leq i \leq n}$ is a random vector in $\mathbb{R}^n$ with dispersion matrix $\boldsymbol{\Sigma}^{n \times n} = (\sigma_{ij})_{1 \leq i \leq n, 1 \leq j \leq n}$:

$$
\sigma_{ij} = \mathrm{Cov}(Y_i, Y_j).
$$

The covariance between any two linear combinations $\sum_i c_i Y_i = \boldsymbol{c}^T \boldsymbol{Y}$ and $\sum_j d_j Y_j = \boldsymbol{d}^T \boldsymbol{Y}$ of the Y's is given by

$$
\mathrm{Cov}(\boldsymbol{c}^T \boldsymbol{Y}, \boldsymbol{d}^T \boldsymbol{Y}) = \sum_{ij} c_i \sigma_{ij} d_j = \boldsymbol{c}^T \boldsymbol{\Sigma} \boldsymbol{d}.
$$

The probabilistic identity

$$
\mathrm{Cov}(\boldsymbol{c}^T \boldsymbol{Y}, \boldsymbol{d}^T \boldsymbol{Y}) = \mathrm{Cov}(\boldsymbol{d}^T \boldsymbol{Y}, \boldsymbol{c}^T \boldsymbol{Y})
$$

implies the algebraic identity

$$\langle c, \Sigma d \rangle = c^T \Sigma d = d^T \Sigma c = \langle \Sigma c, d \rangle.$$

This says that the linear transformation

$$T_\Sigma : x \to \Sigma x$$

is self-adjoint. Notice also that

$$\langle T_\Sigma x, x \rangle = \langle \Sigma x, x \rangle = x^T \Sigma x = \text{Var}(x^T Y).$$

Thus variance/covariance structures are intimately related to self-adjoint transformations.

4.2 Exercise [2]. Let $T: V \to V$ be a linear transformation and let $A = (a_{ik})$ be the matrix of T with respect to a given orthonormal basis $b_1, \ldots, b_n$ for V, so that

$$T b_k = \sum_i a_{ik} b_i \quad \text{for } 1 \le k \le n.$$

Show that

$$a_{jk} = \langle b_j, T b_k \rangle$$

for all j, k and conclude that T is self-adjoint if and only if A is symmetric. ◇

We have seen that any orthogonal projection is self-adjoint. And, of course, the action of an orthogonal projection is geometrically obvious: P_M leaves each $x \in M$ fixed, kills each $y \in M^\perp$, and maps the general V vector $z = x + y$ with $x \in M$ and $y \in M^\perp$ into $P_M z = P_M x + P_M y$. More generally, if

$$V = M_1 + \cdots + M_k$$

is a decomposition of V into mutually orthogonal subspaces and if $\lambda_1, \ldots, \lambda_k$ are any scalars, then

$$T = \sum_{1 \le i \le k} \lambda_i P_{M_i} \qquad (4.3)$$

is self-adjoint, being a linear combination of self-adjoint transformations. Again the action of such a T is geometrically clear: Since

$$T\left(\sum_{1 \le i \le k} x_i \right) = \sum_{1 \le i \le k} \lambda_i x_i \quad \text{for } x_i \in M_i, \ 1 \le i \le k,$$

T just dilates (or contracts) by a factor of λ_i within M_i, $1 \le i \le k$. The remarkable thing is that (4.3) describes the most general self-adjoint transformation on V:

4.4 Theorem (Spectral theorem). *Suppose $T: V \to V$ is linear and self-adjoint. Then there exists an orthogonal decomposition $V = \oplus_{i=1}^k M_i$ of V into nontrivial subspaces M_i and distinct scalars λ_i such that*

$$T = \sum_{1 \le i \le k} \lambda_i P_{M_i}. \qquad (4.5)$$

Suppose T is self-adjoint and therefore of the form (4.5). One can characterize the λ_i's and M_i's in terms of the eigenvalues and eigenvectors of T. Recall that for any linear transformation $S: V \to V$, λ is called an *eigenvalue* of S if there is a nonzero *eigenvector* $x \in V$ such that

$$Sx = \lambda x; \qquad (4.6)$$

the subspace of all $x \in V$ such that (4.6) holds is called the *eigenmanifold* of λ, denoted E_λ. From what was said earlier about the geometric action of T, it is clear that the λ_i's are precisely the distinct eigenvalues of T and the M_i's are the corresponding eigenspaces. Also, for any $x \in V$,

$$Q_T(x) \equiv \langle Tx, x \rangle = \left\langle \sum_i \lambda_i x_i, \sum_j x_j \right\rangle = \sum_i \lambda_i \|x_i\|^2, \qquad (4.7)$$

where $x_i = P_{M_i} x$ for $1 \leq i \leq k$. (The Q in Q_T stands for "quadratic form,") For x of unit length, that is, for

$$1 = \|x\|^2 = \sum_i \|x_i\|^2,$$

$Q_T(x)$ is thus a weighted average of the λ_i's. This observation leads to the following extremal characterization of the λ_i's. Assuming the λ_i's to be indexed in decreasing order, so that $\lambda_1 > \lambda_2 > \cdots > \lambda_k$, one has

$$
\begin{aligned}
\lambda_1 &= \sup\{\, Q_T(x) : \|x\| = 1 \,\} \\
\lambda_2 &= \sup\{\, Q_T(x) : \|x\| = 1,\ x \perp E_{\lambda_1} \,\} \\
&\vdots \qquad\qquad \vdots \\
\lambda_k &= \sup\{\, Q_T(x) : \|x\| = 1,\ x \perp (E_{\lambda_1} + \cdots + E_{\lambda_{k-1}}) \,\}.
\end{aligned}
\qquad (4.8)
$$

Proof of the Spectral theorem. Let T be self-adjoint and linear. We are going to produce the representation (4.5) using (4.8) to define the λ_i's. To begin, consider

$$\lambda_1 \equiv \sup\{\, Q_T(x) : x \in B \,\},$$

where

$$B = \{\, x \in V : \|x\| = 1 \,\}.$$

Since with respect to the distance function

$$d(x, y) = \|y - x\|$$

the closed bounded set B is compact (see Exercise 1.23) and the mapping

$$x \to Q_T(x) = \langle Tx, x \rangle$$

is continuous, and since continuous functions on compact sets achieve their suprema, there exists an $x \in B$ such that

$$Q_T(x) = \lambda_1.$$

We claim that such an x is an eigenvector of T with eigenvalue λ_1, that is,

$$Tx = \lambda_1 x.$$

Notice that

$$\lambda_1 x = \langle Tx, x \rangle x = \frac{\langle Tx, x \rangle}{\langle x, x \rangle}\, x = P_{[x]}(Tx).$$

So if

$$Tx \neq \lambda_1 x,$$

then the residual vector

$$r = Tx - P_{[x]}(Tx)$$

is nonzero, and the idea is to produce a contradiction by perturbing x in the direction r so as to produce a vector $y \in B$ such that $Q_T(y) > Q_T(x)$. Specifically for real δ set

$$y_\delta = \frac{x + \delta r}{\|x + \delta r\|} \in B$$

and note that

$$
\begin{aligned}
Q_T(y_\delta) &= \langle Ty_\delta, y_\delta \rangle \\[4pt]
&= \frac{\langle Tx + \delta Tr, x + \delta r \rangle}{\|x + \delta r\|^2} \\[4pt]
&= \frac{\langle Tx, x \rangle + \delta\big(\langle Tr, x \rangle + \langle Tx, r \rangle\big) + \delta^2 \langle Tr, r \rangle}{\|x + \delta r\|^2} \\[4pt]
&= \frac{\langle Tx, x \rangle + 2\delta \langle Tx, r \rangle + \delta^2 \langle Tr, r \rangle}{\|x + \delta r\|^2} && (T \text{ is self-adjoint}) \\[4pt]
&= \frac{\langle Tx, x \rangle + 2\delta\|r\|^2 + \delta^2 \langle Tr, r \rangle}{\|x\|^2 + \delta^2\|r\|^2} && (r = Q_{[x]}Tx \perp x).
\end{aligned}
$$

Since $\|r\| \neq 0$ by supposition, $Q_T(y_\delta) > \langle Tx, x \rangle / \|x\|^2 = Q_T(x)$ for sufficiently small positive δ.

This shows that λ_1 is an eigenvalue of T. Let M_1 be the associated eigenspace. If $M_1 = V$, then $T = \lambda_1 P_{M_1}$ and we are done. Otherwise note that for $x \in M_1$ and $y \in M_1^\perp$,

$$\langle x, Ty \rangle = \langle Tx, y \rangle = \lambda_1 \langle x, y \rangle = 0,$$

so T maps $M_1^\perp$ into itself. Applying the preceding argument to the restriction T_1 of T to $M_1^\perp$ we find that

$$\lambda_2 = \sup\{\, Q_{T_1}(x) \equiv Q_T(x) : x \in B \text{ and } x \perp M_1 \,\}$$

is an eigenvalue of T_1 (and therefore of T) with some nontrivial eigenspace $M_2 \subset M_1^\perp$. If $M_2 = M_1^\perp$, we have $T = \lambda_1 P_{M_1} + \lambda_2 P_{M_2}$ and are done. Otherwise we recursively define λ_i and M_i for $i = 3, \ldots$ by

$$\lambda_i = \sup\{\, Q_T(x) : x \in B \text{ and } x \perp (M_1 + \cdots + M_{i-1}) \,\}$$
$$M_i = \text{eigenspace of } \lambda_i \text{ for } T \text{ restricted to } (M_1 + \cdots + M_{i-1})^\perp,$$

stopping the process with the first index k such that $\sum_{1 \le i \le k} M_i$ is V; such a k exists because each new eigenspace adds at least one dimension to the sum. We then have $T = \sum_{1 \le i \le k} \lambda_i P_{M_i}$. ∎

4.9 Exercise [3]. Find the spectral representation (4.5) of the transformation $T \colon \mathbb{R}^n \to \mathbb{R}^n$ corresponding to the matrix

$$\begin{pmatrix} a & b & b & \cdots & b \\ b & a & b & \cdots & b \\ b & b & a & \cdots & b \\ \vdots & \vdots & \vdots & \cdots & \vdots \\ b & b & b & \cdots & a \end{pmatrix}.$$ ◇

4.10 Exercise [2]. Use the Spectral theorem to solve Exercise 2.7. ◇

4.11 Exercise [3]. Use the Spectral theorem to show that if $A^{n \times n}$ is a symmetric matrix, then there exists an orthogonal matrix $\mathcal{O}^{n \times n}$ such that $D \equiv \mathcal{O}^T A \mathcal{O}$ is diagonal. ◇

4.12 Exercise [3]. Suppose $T \colon V \to V$ is linear self-adjoint and positive semi-definite ($\langle v, Tv \rangle \ge 0$ for all $v \in V$). Show that there exists a unique linear self-adjoint positive semi-definite transformation $S \colon V \to V$ such that $T = S^2$. (S is called the (positive) *square root* of T; one writes $S = \sqrt{T}$.) ◇

5. Representation of linear and bilinear functionals

For fixed $v \in V$, the mapping

$$x \to \langle x, v \rangle \quad (x \in V)$$

is a *linear functional* on V, that is, a linear transformation from V to $\mathbb{R}$. The following result says every linear functional on V is of this form.

5.1 Proposition (Representation theorem for linear functionals). *If ψ is a linear functional on V, then there exists a unique $v \in V$ such that*

$$\psi(x) = \langle x, v \rangle \quad \text{for all } x \in V. \tag{5.2}$$

The v of (5.2) is called the *coefficient vector* of ψ; one writes $v = c.v.(\psi)$, or just $v = cv(\psi)$.

Proof of the Representation theorem. *Uniqueness:* If v and w are both coefficient vectors for ψ, then their difference is orthogonal to every vector in V and so is 0.

Existence: If the desired representation holds, then the null space of ψ is the orthogonal complement of $[v]$ in V. This observation suggests how to proceed. To avoid trivialities, suppose ψ is not identically 0 and let N be its null space. Since

$$d(N) = d(V) - d\big(\mathcal{R}(\psi)\big) = d(V) - 1,$$

$N^{\perp}$ is one-dimensional and so of the form $[w]$ for some $0 \neq w \in V$. The idea is to exhibit the desired coefficient vector as a scalar multiple of w, say dw. Since the general vector

$$z = y + cw \qquad (y \in N, \ c \in \mathbb{R})$$

of V is mapped by ψ into

$$\psi(z) = \psi(y) + c\psi(w) = c\psi(w)$$

and by $\langle \cdot, dw \rangle$ into

$$\langle z, dw \rangle = \langle y + cw, dw \rangle = cd\langle w, w \rangle,$$

the appropriate choice of d is

$$d = \frac{\psi(w)}{\|w\|^2} \, . \qquad\blacksquare$$

There are two common ways to produce the coefficient vector v of a given linear functional. One way is the *direct-construction method*, which constructs v as in the existence part of the proof. The other way is the *confirmation method*, which checks that a conjectured v satisfies (5.2) and then appeals to uniqueness.

5.3 Exercise [3]. Suppose $x_1, x_2, \ldots, x_n$ is a basis for V, so that each $x \in V$ has a unique representation of the form

$$x = x_1\beta_1 + x_2\beta_2 + \cdots + x_n\beta_n.$$

Show that the coefficient vector of the so-called j^{th}-*coordinate functional*

$$\psi_j : x \to \beta_j$$

is

$$v_j \equiv \left(\frac{Q_j x_j}{\|Q_j x_j\|} \right) \frac{1}{\|Q_j x_j\|}, \tag{5.4}$$

where $Q_j = I - P_j$ with P_j being projection onto the manifold spanned by the x_i's for $i \neq j$. Solve this exercise twice, once using the direct-construction method and once using the confirmation method. $\diamond$

Suppose now $F : V \times V \to \mathbb{R}$ is a *bilinear functional*, that is, linear in each component, the other component being held fixed. For each $z \in V$, the map

$$y \to F(y, z)$$

is linear and so of the form

$$F(y, z) = \langle y, Cz \rangle$$

for a unique vector $Cz \in V$. Since

$$\langle y, C(c_1 z_1 + c_2 z_2) \rangle = F(y, c_1 z_1 + c_2 z_2) = c_1 F(y, z_1) + c_2 F(y, z_2)$$
$$= \langle y, c_1 C z_1 + c_2 C z_2 \rangle$$

for all $y \in V$, the mapping $C \colon V \to V$ is linear. We have established:

5.5 Proposition (Representation theorem for bilinear functionals). *If $F \colon V \times V \to \mathbb{R}$ is bilinear, then there exists a unique linear transformation $C \colon V \to V$ such that*

$$F(x, y) = \langle x, Cy \rangle \quad \text{for all } x \text{ and } y \in V. \tag{5.6}$$

This result has several important consequences. First, if $T \colon V \to V$ is linear, then

$$F(x, y) \equiv \langle Tx, y \rangle$$

defines a bilinear functional on V that is necessarily of the form

$$\langle Tx, y \rangle = \langle x, T'y \rangle$$

for some unique linear transformation $T' \colon V \to V$. T' is called the *adjoint* of T. One has

$$(T')' = T, \quad (ST)' = T'S', \tag{5.7}$$

and $T' = T$ if and only if T is self-adjoint in the sense of (4.1). Moreover, T is orthogonal if and only if $T' = T^{-1}$, because the identity

$$\langle Tx, Ty \rangle - \langle x, y \rangle = \langle (T'T - I)x, y \rangle$$

for $x, y \in V$ implies that T preserves inner products if and only if $T'T = I$.

5.8 Exercise [1]. Let T be a linear transformation from V to V, and let $\boldsymbol{A} = (a_{ij})$ be the matrix of T with respect to a given orthonormal basis for V. Show that the matrix of T' is the transpose (a_{ji}) of $\boldsymbol{A}$. ◇

5.9 Exercise [3]. Let T be a linear transformation from V to V. Show that $\mathcal{N}(T') = (\mathcal{R}(T))^{\perp}$ and deduce that T' is nonsingular if and only if T is. ◇

5.10 Exercise [3]. Let M be a subspace of V and let $\mathcal{O} \colon V \to V$ be an orthogonal linear transformation. Show that $\mathcal{O}P_M \mathcal{O}'$ is orthogonal projection onto $\mathcal{O}(M)$. ◇

5.11 Exercise [3]. Locate at least two places where use of the relation $(ST)' = T'S'$ would simplify the calculations in Section 2.2. ◇

Next suppose that $F \colon V \times V \to \mathbb{R}$ is both bilinear and symmetric. The representing transformation C is then self-adjoint and one can find an orthonormal basis for V consisting of the eigenvectors $b_1, \ldots, b_n$ of C with the associated eigenvalues $\lambda_1, \ldots, \lambda_n$. F then takes the form

$$F(x, y) = \langle x, Cy \rangle = \left\langle \sum_i c_i b_i, C \sum_j d_j b_j \right\rangle = \sum_i \lambda_i c_i d_i, \tag{5.12}$$

where the $c_i = \langle x, b_i \rangle$'s and $d_j = \langle y, b_j \rangle$'s are the coefficients of x and y with respect to the b-basis. Thus, when viewed from the right perspective, the general symmetric bilinear functional on V is just a weighted dot-product (the weights can be zero or negative).

5.13 Exercise [3]. A *quadratic form* Q on V is a functional of the form

$$Q(v) = \langle Cv, v \rangle$$

for some linear transformation $C: V \to V$. Show that there is no loss of generality in supposing that C is self-adjoint, in which case C is uniquely determined by Q. [Hint: For uniqueness, generalize (1.6) to

$$\langle Cv, w \rangle = \frac{Q(v + w) - Q(v) - Q(w)}{2} .] \tag{5.14} \diamond$$

Finally, suppose $F = [\cdot, \cdot]$ is another inner product on V, so that F is positive-definite in addition to being symmetric and bilinear. We then have

$$[x, y] = \langle x, Cy \rangle \quad \text{for all } x \text{ and } y \text{ in } V \tag{5.15}$$

for a unique self-adjoint linear transformation C on V that is positive-definite in the sense that

$$\langle x, Cx \rangle \geq 0 \text{ for all } x \in V \quad \text{and} \quad \langle x, Cx \rangle = 0 \text{ only for } x = 0.$$

In other words, with respect to an appropriate orthonormal basis $b_1, \ldots, b_n$, $[\cdot, \cdot]$ takes the form

$$[x, y] = \sum_i \lambda_i c_i d_i, \quad \text{for } c_i = \langle x, b_i \rangle, \ d_i = \langle y, b_i \rangle, \tag{5.16}$$

with the weights $\lambda_1, \lambda_2, \ldots, \lambda_n$ being strictly positive.

5.17 Exercise [4]. Suppose V and W are both inner product spaces. Formulate and prove a representation theorem for bilinear functionals on $V \times W$. Show that if $T: V \to W$ is linear, then there exists a unique linear transformation $T': W \to V$ such that

$$\langle Tv, w \rangle_W = \langle v, T'w \rangle_V \quad \text{for all } v \in V \text{ and } w \in W; \tag{5.18}$$

T' is called the *adjoint of* T. Formulate and prove extensions of (5.7) and the assertion of Exercise 5.8. $\diamond$

5.19 Exercise [2]. Suppose M is a subspace of V. Up to now we have considered P_M as a linear transformation from V to V, and as such it is self-adjoint. One can, however, think of P_M as being a linear transformation from V to M. If one takes this point of view, what then is the adjoint of P_M? (The inner product of two elements of M is defined to be their inner product in V.) $\diamond$

5.20 Exercise [3]. Suppose V and W are both inner product spaces and that $T: V \to W$ is linear. Show that

$$\mathcal{R}(T') = \mathcal{R}(T'T) \quad \text{and} \quad \mathcal{N}(T) = \mathcal{N}(T'T). \tag{5.21}$$

Deduce that $T'T$ is nonsingular if and only if T is. ◇

5.22 Exercise [3]. Suppose B and V are both inner product spaces and that $T: B \to V$ is linear. Put $M = T(B) = \{ Tb : b \in B \}$. Show that M is a subspace of V and that for each $v \in V$, one has $P_M v = Tb$ if and only if $b \in B$ satisfies the so-called *normal equations* $T'Tb = Tv$. Do not assume that T is nonsingular. ◇

5.23 Exercise [3]. Suppose $\langle \cdot, \cdot \rangle$ and $[\cdot, \cdot]$ are both inner products on V. Express the unique $[\cdot, \cdot]$-self-adjoint, positive-definite linear transformation D such that $\langle x, y \rangle = [x, Dy]$ for all $x, y \in V$ in terms of the C of equation (5.15). ◇

6. Problem set: Cleveland's identity

Let $(V, \langle \cdot, \cdot \rangle)$ be an inner product space. Let $\langle \cdot, \cdot \rangle^* = \langle \cdot, \Delta \cdot \rangle$ be another inner product on V: Δ is a positive-definite self-adjoint linear transformation from V to V. For any subspace M of V, let P_M denote projection onto M with respect to $\langle \cdot, \cdot \rangle$, and let P_M^* denote projection onto M with respect to $\langle \cdot, \cdot \rangle^*$. Denote the corresponding residual projections by Q_M and Q_M^*, respectively. For any vector z in V, one has of course

$$\|Q_M^* z\|^2 \geq \|Q_M z\|^2.$$

The question of how much larger can $\|Q_M^* z\|^2$ be than $\|Q_M z\|^2$ is addressed by *Cleveland's identity*, which asserts that

$$\sup_M \sup_{z \in V, z \notin M} \frac{\|Q_M^* z\|^2}{\|Q_M z\|^2} = \frac{(1+\tau)^2}{4\tau}, \tag{6.1}$$

where

$$\tau = \frac{\lambda_n}{\lambda_1} \tag{6.2}$$

is the ratio of the largest eigenvalue λ_n of Δ to the smallest eigenvalue λ_1 of Δ. By the extremal characterization of the eigenvalues of Δ, one has

$$\lambda_1 = \inf_{v \in V, v \neq 0} \left(\frac{\|v\|^*}{\|v\|} \right)^2 \quad \text{and} \quad \lambda_n = \sup_{v \in V, v \neq 0} \left(\frac{\|v\|^*}{\|v\|} \right)^2. \tag{6.3}$$

In this problem set you are asked to establish (6.1). Here are the steps in the argument. To begin with, suppose M and $z \notin M$ are given; one needs to show

$$\frac{\|Q_M^* z\|^2}{\|Q_M z\|^2} \leq \frac{(1+\tau)^2}{4\tau}. \tag{6.4}$$

A. Show that, without loss of generality, one can consider just the case where $z \perp M$ (with respect to $\langle \cdot, \cdot \rangle$) with $\|z\| = 1$ and where M is one-dimensional, say $M = [y]$

with $\|y\| = 1$.

[Hint: Given z and M such that $P_M z \neq P_M^* z$, find a one-dimensional subspace L of M such that

$$\|Q_M^* z\| = \|Q_L^* u\| \quad \text{and} \quad \|Q_M z\| = \|Q_L u\|$$

for $u = Q_M z$.]

B. Let y and z be as in part **A**. Put $W = [y, z]$. Show that there exists a basis for W, say $\{b_1, b_2\}$, orthonormal with respect to $\langle \cdot, \cdot \rangle$, such that, for $v, w \in W$,

$$\langle v, w \rangle^* = \beta_1 v_1 w_1 + \beta_2 v_2 w_2,$$

where $v_i = \langle v, b_i \rangle$, $w_i = \langle w, b_i \rangle$, and

$$\beta_1 = \inf_{u \in W, u \neq 0} \left(\frac{\|u\|^*}{\|u\|} \right)^2 \quad \text{and} \quad \beta_2 = \sup_{u \in W, u \neq 0} \left(\frac{\|u\|^*}{\|u\|} \right)^2.$$

[Hint: Restrict $\langle \cdot, \cdot \rangle$ and $\langle \cdot, \cdot \rangle^*$ to $W \times W$.]

C. Let y and z be as in part **A** and b_1, b_2, β_1, β_2 as in part **B**. Show that

$$\frac{\|Q_M^* z\|^2}{\|Q_M z\|^2} = \frac{\beta_1^2 y_1^2 + \beta_2^2 y_2^2}{(\beta_1 y_1^2 + \beta_2 y_2^2)^2},$$

where $y_i = \langle y, b_i \rangle$.

[Hint: Compute! Note that the matrix $\begin{pmatrix} y_1 & z_1 \\ y_2 & z_2 \end{pmatrix}$ is orthogonal.]

D. Show that

$$\sup_{p_1, p_2 \geq 0; \, p_1 + p_2 = 1} \frac{\beta_1^2 p_1 + \beta_2^2 p_2}{(\beta_1 p_1 + \beta_2 p_2)^2} = \frac{(\beta_1 + \beta_2)^2}{4\beta_1 \beta_2}$$

and deduce (6.4).

E. Show that equality holds in (6.1) by exhibiting M and z such that

$$\frac{\|Q_M^* z\|^2}{\|Q_M z\|^2} = \frac{(1 + \tau)^2}{4\tau}.$$

7. Appendix: Rudiments

This appendix reviews basic definitions and facts about linear algebra with which we presume the reader is acquainted.

7A. Vector spaces

A (real) *vector space* is a triple $(V, +, \cdot)$ consisting of a set V of objects called vectors and two operations, vector addition, $+$, which associates to each pair v_1, v_2 of vectors in V a vector $v_1 + v_2$ in V, and scalar multiplication, $\cdot$, which associates to each vector $v \in V$ and each scalar $c \in \mathbb{R}$ a vector $c \cdot v \equiv cv$ in V. The operations are assumed to have the following properties:

(1) For all $v_1, v_2, v_3 \in V$, $(v_1 + v_2) + v_3 = v_1 + (v_2 + v_3)$.

(2) For all $v_1, v_2 \in V$, $v_1 + v_2 = v_2 + v_1$.

(3) There exists a vector $0 \in V$ such that $v + 0 = v$ for all $v \in V$.

(4) For each $v \in V$, there exists a vector $-v \in V$ such that $-v + v = 0$. For $v_1 + (-v_2)$ we write also $v_1 - v_2$.

(5) For all $c \in \mathbb{R}$ and $v_1, v_2 \in V$, $c(v_1 + v_2) = cv_1 + cv_2$.

(6) For all $c_1, c_2 \in \mathbb{R}$ and $v \in V$, $(c_1 + c_2)v = c_1 v + c_2 v$.

(7) For all $c_1, c_2 \in \mathbb{R}$ and $v \in V$, $(c_1 c_2)v = c_1(c_2 v)$.

(8) For all $v \in V$, $1 \cdot v = v$.

An important example of a vector space is $\mathbb{R}^n$, the set of n-tuples (or $n \times 1$ column vectors) of real numbers, for which vector addition and multiplication by scalars are defined componentwise as addition and multiplication of real numbers. The *transpose* of a column vector $\boldsymbol{x} \in \mathbb{R}^n$ with components $x_1, \ldots, x_n$ is the row vector $\boldsymbol{x}^T = (x_1, \ldots, x_n)$.

Vectors $v_1, \ldots, v_n$ in a vector space V are said to be *(linearly) independent* if

$$\sum_{1 \le i \le n} c_i v_i = 0$$

implies that the c_i's are all zero; otherwise the v_i's are said to be *(linearly) dependent*. The subset of M of vectors in V of the form $\sum_{1 \le i \le n} c_i v_i$, where each c_i varies over $\mathbb{R}$, is called the *span* of $v_1, \ldots, v_n$, denoted $[v_1, \ldots, v_n]$. Vectors $v_1, \ldots, v_n$ are said to form a *basis* for V if and only if they are linearly independent and V is their span, that is, if and only if each vector in V has a unique representation of the form $\sum_i c_i v_i$. When $v_1, \ldots, v_n$ is a basis for V, c_i is called the *coordinate of* $\sum_{1 \le j \le n} c_j v_j$ *with respect to* v_i.

V is said to be *finite-dimensional* if it has a basis consisting of finitely many elements. Every basis for a finite-dimensional vector space has the same number of elements; this number is called the *dimension* of V, denoted $\dim(V)$, or just $d(V)$. If V is an n-dimensional vector space, then $v_1, \ldots, v_n$ is a basis for V if and only if the v_i's are linearly independent or, equivalently, if and only if they span V. The *usual coordinate basis* for $\mathbb{R}^n$ is comprised of the vectors $e^{(1)}, \ldots, e^{(n)}$ defined by $e_j^{(i)} = \delta_{ij}$.

We assume henceforth that all vector spaces we deal with are finite-dimensional.

7B. Subspaces

A nonempty subset M of a vector space V that is closed under vector addition and multiplication by scalars is called a *subspace*, or *(linear) manifold*, of V. For example, the span $[v_1, \ldots, v_m]$ of given vectors $v_1, \ldots, v_m$ is a subspace. We write the trivial subspace $[0]$ simply as 0. A subset F of V of the form $F = v_0 + M = \{\, v_0 + m : m \in M \,\}$, where M is a subspace and $v_0 \in V$, is called a *flat*, or *shifted manifold*; F is a flat if and only if $c_1 f_1 + c_2 f_2 \in F$ whenever $f_1, f_2 \in F$ and $c_1 + c_2 = 1$. The dimension of F is defined to be the dimension of M.

Two subspaces of V are said to be *disjoint* if they have only the zero vector in common (this notion should not be confused with the set-theoretic one). The *intersection* $M_1 \cap M_2$ of two subspaces is defined to be the subspace of V consisting of all vectors common to M_1 and to M_2. To say $M_1 \cap M_2 = 0$ is to say M_1 and M_2 are disjoint. The *sum* $M_1 + M_2$ of two subspaces of V is defined to be the subspace of V consisting of all vectors of the form $m_1 + m_2$, where $m_1 \in M_1$ and $m_2 \in M_2$. A sum $M_1 + M_2$ is said to be a *direct sum*, denoted $M_1 \oplus M_2$, if M_1 and M_2 are disjoint, that is, if the representation of $m \in M_1 + M_2$ as $m = m_1 + m_2$ with $m_i \in M_i$ is unique. As regards to dimension, one has

$$d(M_1 + M_2) + d(M_1 \cap M_2) = d(M_1) + d(M_2); \tag{7.1}$$

in particular, $d(M_1 \oplus M_2) = d(M_1) + d(M_2)$.

7C. Linear functionals

A mapping ψ of V into $\mathbb{R}$ is called a *linear functional* on V provided it preserves addition and scalar multiplication: $\psi(c_1 v_1 + c_2 v_2) = c_1 \psi(v_1) + c_2 \psi(v_2)$ for all $c_1, c_2 \in \mathbb{R}$ and $v_1, v_2 \in V$. Operations of addition and scalar multiplication on the set V^* of all linear functionals on V are defined pointwise, so that $(\psi_1 + \psi_2)(v) = \psi_1(v) + \psi_2(v)$ and $(c\psi)(v) = c\psi(v)$. Under these operations, V^* is itself a vector space, called the *dual* of V. V^* has the same dimension as V. Examples of linear functionals on V are the *coordinate functionals* ψ_i defined relative to a given basis $v_1, \ldots, v_n$ for V by $\psi_i(\sum_{1 \le j \le n} c_j v_j) = c_i$ for $i = 1, \ldots, n$.

7D. Linear transformations

A mapping T of a vector space V into a vector space W is called a *linear transformation* if it preserves linear structure: $T(c_1 v_1 + c_2 v_2) = c_1 T(v_1) + c_2 T(v_2)$ for all $c_1, c_2 \in \mathbb{R}$ and $v_1, v_2 \in V$. Addition and scalar multiplication of linear transformations from V to W are defined pointwise. A linear functional is simply a linear transformation from V into $\mathbb{R}^1$. A linear transformation $T : V \to W$ is called a *isomorphism* if it is one-to-one and onto. V and W are said to be *isomorphic* if there is an isomorphism mapping one onto the other; isomorphic vector spaces are identical so far as their linear structure is concerned. The *composition* of the linear transformations $S : U \to V$ and $T : V \to W$ is defined to be the linear transformation $TS : U \to W$ that sends $u \in U$ to $T(S(u)) \in W$. A linear transformation $T : V \to V$ such that $T = T^2 \ (\equiv TT)$ is said to be *idempotent*. The linear transformation that leaves each $v \in V$ fixed is called the *identity* transformation and is commonly denoted by I or I_V.

The *range* of a linear transformation T mapping V to W is the subspace of W, denoted $\mathcal{R}(T)$ or $R(T)$, consisting of all vectors of the form Tv for $v \in V$. The *rank*, $\rho(T)$, of T is the dimension of $\mathcal{R}(T)$. The *null space* of T is the subspace of V, denoted $\mathcal{N}(T)$ or $N(T)$, consisting of all vectors in V that are mapped by T into zero in W.

The formula

$$d\big(\mathcal{R}(T)\big) + d\big(\mathcal{N}(T)\big) = d(V), \tag{7.2}$$

holding for any linear transformation T from V to W, has many applications. In particular, if V and W have the same dimension and $\mathcal{N}(T) = 0$, then T is an isomorphism of V and W. A transformation T such that $\mathcal{N}(T) = 0$ is said to be *nonsingular*. The *inverse*, T^{-1}, of a nonsingular transformation exists on $\mathcal{R}(T)$ and is linear.

The *matrix* of a linear transformation $T\colon V \to W$ with respect to bases $v_1, \ldots, v_m$ for V and $w_1, \ldots, w_n$ for W is the $n \times m$ array $[T] \equiv (t_{ij})$ defined by $T(v_j) = \sum_{1 \le i \le n} t_{ij} w_i$ for $1 \le j \le m$ or, equivalently, by the condition $T(\sum_j c_j v_j) = \sum_i (\sum_j t_{ij} c_j) w_i$ for $c_1, \ldots, c_m \in V$. If $S\colon U \to V$ and $T\colon V \to W$ are both linear, then the identity $[TS] = [T][S]$ holds relative to fixed bases for U, V, and W.

The *transpose* of an $m \times n$ matrix $\boldsymbol{A} = (a_{ij})_{i=1,\ldots,m,\,j=1,\ldots n}$ is the $n \times m$ matrix $\boldsymbol{A}^T = (b_{ji})_{j=1,\ldots,n,\,i=1,\ldots,m}$, where $b_{ji} = a_{ij}$.

CHAPTER 3

RANDOM VECTORS

Here we develop certain aspects of the concept of a random vector, say Y, taking values in an inner product space. Particular attention is paid to the notions of weak sphericity and normality. The theory is based on linear functionals of Y. Most of this material constitutes a generalization of the elementary properties of random vectors in $\mathbb{R}^n$, with which it is assumed that the reader is familiar.

Throughout $(V, \langle \cdot, \cdot \rangle)$ is an inner product space and all indicated expectations of real random variables are assumed to exist and be finite, unless specifically noted to the contrary. Recall that, by convention, we always view $\mathbb{R}^n$ as being endowed with the dot-product, unless specifically noted to the contrary.

1. Random vectors taking values in an inner product space

Motivation. Let $Y_1, \ldots, Y_n$ be random variables and set $\boldsymbol{Y} = (Y_1, \ldots, Y_n)^T$. $\boldsymbol{Y}$ takes values in $\mathbb{R}^n$. For any constant vector $\boldsymbol{c} = (c_1, \ldots, c_n)^T \in \mathbb{R}^n$,

$$\langle \boldsymbol{c}, \boldsymbol{Y} \rangle = \sum_i c_i Y_i \tag{1.1}$$

is a random variable. Conversely, if one starts with the assumption that $\langle \boldsymbol{c}, \boldsymbol{Y} \rangle$ is a random variable for all $\boldsymbol{c} \in \mathbb{R}^n$, then it follows that each component of $\boldsymbol{Y}$ is a random variable; indeed, for $1 \le j \le n$ one has $Y_j = \langle \boldsymbol{e}^{(j)}, \boldsymbol{Y} \rangle$, where

$$(\boldsymbol{e}^{(j)})_i = \begin{cases} 1, & \text{if } i = j, \\ 0, & \text{if } i \ne j. \end{cases} \tag{1.2}$$

These observations lead to the definition made below in the coordinate-free case. $\diamond$

Let Y be a function defined on some probability space and taking values in V. Y is said to be a V-*valued random vector* if $\langle v, Y \rangle$ is a random variable for all $v \in V$, that is, if every linear functional of Y is a random variable. An equivalent requirement is that Y be measurable with respect to the smallest σ-field $\mathcal{B}$ rendering measurable all linear functionals on V. $\mathcal{B}$ turns out to be identical to the so-called Borel σ-field of V, generated by the open sets of V in the natural metric $(d(x, y) =$

$\|y - x\|$); for the most part, we shall brush aside measure-theoretic considerations such as these.

1.3 Exercise [4]. You should take note of the assertions of this exercise even if you do not have at your command the (elementary) properties of measurable functions that are needed to prove them. Show that: (1) Y is a random vector in V if and only if $\langle b_i, Y \rangle$ is a random variable for each b_i in some fixed basis $b_1, \ldots, b_n$ for V; (2) if Y is a V-valued random vector, then $f(Y)$ is a real random variable for every continuous function $f: V \to \mathbb{R}$.   ◇

2. Expected values

Motivation. Return again to $\mathbb{R}^n$. It is customary to define the expected value of an $\mathbb{R}^n$-valued random vector $\boldsymbol{Y} = (Y_1, \ldots, Y_n)^T$ to be the vector $\boldsymbol{\mu} = (EY_1, \ldots, EY_n)^T$. For each $\boldsymbol{c} \in \mathbb{R}^n$ one then has

$$E(\langle \boldsymbol{c}, \boldsymbol{Y} \rangle) = E\left(\sum_i c_i Y_i\right) = \sum_i c_i EY_i = \langle \boldsymbol{c}, \boldsymbol{\mu} \rangle. \tag{2.1}$$

Moreover, $\boldsymbol{\mu} \equiv (\mu_1, \ldots, \mu_n)^T$ is uniquely determined by these relations: Taking $\boldsymbol{c} = \boldsymbol{e}^{(j)}$ in (2.1) gives

$$\mu_j = \langle \boldsymbol{e}^{(j)}, \boldsymbol{\mu} \rangle = E(\langle \boldsymbol{e}^{(j)}, \boldsymbol{Y} \rangle) = E(Y_j).   ◇$$

In the coordinate-free setting, one adopts an analog of (2.1) as the definition, as follows. Suppose Y is a random vector in V. The map

$$v \to E(\langle v, Y \rangle)$$

is a linear functional on V, and so there exists a vector, say EY, unique with respect to $\langle \cdot, \cdot \rangle$, such that

$$E(\langle v, Y \rangle) = \langle v, EY \rangle \quad \text{for all } v \in V. \tag{2.2}$$

EY is called the *expectation*, or *mean*, of Y. Note that formula (2.2) says $E(\psi(Y)) = \psi(EY)$ for every linear functional ψ on V. It follows that EY does not depend on the inner product employed on V.

2.3 Exercise [2]. Show that a V-valued random variable Y *has* an expectation, in the sense that the left-hand side of (2.2) is defined and finite for all $v \in V$, if and only if $E(\|Y\|) < \infty$.
[Hint: Show $E(|\langle v, Y \rangle|) < \infty$ for all $v \in V$ if and only if $E(\|Y\|) < \infty$.]   ◇

Expectation is preserved under linear transformation: If $T: V \to V$ is linear, then

$$E(TY) = T(EY) \tag{2.4}$$

because

$$E(\langle v, TY \rangle) = E(\langle T'v, Y \rangle) = \langle T'v, EY \rangle = \langle v, TEY \rangle \quad \text{for all } v \in V.$$

Moreover, expectation is a linear operation: If Y_1 and Y_2 are V-valued random vectors and c_1 and c_2 are real numbers, then

$$E(c_1 Y_1 + c_2 Y_2) = c_1 E(Y_1) + c_2 E(Y_2) \tag{2.5}$$

because

$$E(\langle v, c_1 Y_1 + c_2 Y_2 \rangle) = c_1 E\langle v, Y_1 \rangle + c_2 E\langle v, Y_2 \rangle = c_1 \langle v, EY_1 \rangle + c_2 \langle v, EY_2 \rangle$$
$$= \langle v, c_1 EY_1 + c_2 EY_2 \rangle$$

for all $v \in V$.

2.6 Exercise [1]. Show that if X is a real random variable and $w \in V$, then $E(Xw) = E(X)w$. ◇

2.7 Exercise [1]. Show that if Y is a V-valued random vector that takes all its values in a subspace M of V, then $EY \in M$. ◇

3. Covariance operators

Motivation. Let $\boldsymbol{Y} = (Y_1, \ldots, Y_n)^T$ and $\boldsymbol{Z} = (Z_1, \ldots, Z_n)^T$ be random vectors in $\mathbb{R}^n$. The *covariance matrix* of $\boldsymbol{Y}$ and $\boldsymbol{Z}$ is the $n \times n$ matrix

$$\boldsymbol{C} = (c_{ij}) = \big(\mathrm{Cov}(Y_i, Z_j)\big)_{1 \le i \le n, 1 \le j \le n}. \tag{3.1}$$

Knowing $\boldsymbol{C}$, one may compute the covariance between any linear combination of the Y_i's and any linear combination of the Z_j's:

$$\mathrm{Cov}\big(\langle \boldsymbol{b}, \boldsymbol{Y} \rangle, \langle \boldsymbol{d}, \boldsymbol{Z} \rangle\big) = \mathrm{Cov}\Big(\sum_i b_i Y_i, \sum_j d_j Z_j\Big)$$
$$= \sum_{ij} b_i c_{ij} d_j = \sum_i b_i \Big(\sum_j c_{ij} d_j\Big) = \langle \boldsymbol{b}, \boldsymbol{C} \boldsymbol{d} \rangle \tag{3.2}$$

for all $\boldsymbol{b} \in \mathbb{R}^n$ and $\boldsymbol{d} \in \mathbb{R}^n$. Moreover, $\boldsymbol{C}$ is uniquely determined by these relations because

$$c_{ij} = \langle \boldsymbol{e}^{(i)}, \boldsymbol{C} \boldsymbol{e}^{(j)} \rangle = \mathrm{Cov}\big(\langle \boldsymbol{e}^{(i)}, \boldsymbol{Y} \rangle, \langle \boldsymbol{e}^{(j)}, \boldsymbol{Z} \rangle\big) = \mathrm{Cov}(Y_i, Z_j). \quad ◇$$

In the coordinate-free setting, an analog of (3.2) is taken as the definition. For this suppose Y and Z are V-valued random vectors. Set

$$F(y, z) = \mathrm{Cov}\big(\langle y, Y \rangle, \langle z, Z \rangle\big) \quad \text{for } y, z \in V.$$

Since F is a bilinear functional on V, there exists a unique linear transformation C from V to V such that

$$\mathrm{Cov}\big(\langle y, Y \rangle, \langle z, Z \rangle\big) = \langle y, Cz \rangle \quad \text{for all } y, z \in V. \tag{3.3}$$

C is called the *covariance operator of Y and Z*, denoted $\mathrm{Cov}(Y, Z)$. If $\boldsymbol{Y}$ and $\boldsymbol{Z}$ are $\mathbb{R}^n$-valued random vectors, then, according to the preceding discussion, the covariance operator of $\boldsymbol{Y}$ and $\boldsymbol{Z}$ is the transformation having matrix (3.1) with respect to the standard basis $\boldsymbol{e}^{(1)}, \ldots, \boldsymbol{e}^{(n)}$.

It is important to note that $\mathrm{Cov}(Y, Z)$ *does* depend on the inner product employed on V. Indeed, if

$$\langle \cdot, \cdot \rangle_\Delta = \langle \Delta \cdot, \cdot \rangle$$

is another inner product on V, with $\Delta \colon V \to V$ being linear, self-adjoint, and positive-definite (see (2.5.15)), and if C satisfies (3.3), then

$$\mathrm{Cov}\big(\langle y, Y \rangle_\Delta, \langle z, Z \rangle_\Delta\big) = \mathrm{Cov}\big(\langle \Delta y, Y \rangle, \langle \Delta z, Z \rangle\big) = \langle \Delta y, C \Delta z \rangle = \langle y, C \Delta z \rangle_\Delta$$

for all $y, z \in V$, so

$$\begin{array}{cc} C\Delta \text{ is the covariance operator of } Y \text{ and } Z \\ \text{in the } \langle \cdot, \cdot \rangle_\Delta \text{ inner product.} \end{array} \tag{3.4}$$

Here are two useful properties of covariance operators:

$$\mathrm{Cov}(Z, Y) = \big(\mathrm{Cov}(Y, Z)\big)', \tag{3.5}$$

$$\mathrm{Cov}(TY, UZ) = T \, \mathrm{Cov}(Y, Z) U' \quad \text{for linear } T, U \colon V \to V. \tag{3.6}$$

The proofs are elementary:

$$\begin{aligned} \mathrm{Cov}\big(\langle y, Z \rangle, \langle z, Y \rangle\big) &= \mathrm{Cov}\big(\langle z, Y \rangle, \langle y, Z \rangle\big) \\ &= \langle z, Cy \rangle = \langle Cy, z \rangle = \langle y, C'z \rangle \end{aligned}$$

implies $\mathrm{Cov}(Z, Y) = C'$, with C being $\mathrm{Cov}(Y, Z)$, and

$$\begin{aligned} \mathrm{Cov}\big(\langle y, TY \rangle, \langle z, UZ \rangle\big) &= \mathrm{Cov}\big(\langle T'y, Y \rangle, \langle U'z, Z \rangle\big) \\ &= \langle T'y, CU'z \rangle = \langle y, TCU'z \rangle \end{aligned}$$

implies $\mathrm{Cov}(TY, UZ) = T \, \mathrm{Cov}(Y, Z) U'$.

V-valued random vectors $Y_1, \ldots, Y_k$ are said to be *uncorrelated* if $\mathrm{Cov}(Y_i, Y_j) = 0$ for $1 \leq i < j \leq k$; by (3.4), this notion does not depend on the inner product employed.

3.7 Exercise [3]. Generalize (2.4), (3.3), (3.4), (3.5), and (3.6) to the case where Y and/or Z, T, and/or U do not necessarily take values in the same spaces (formulate and prove the appropriate extensions). Make use of the notion of adjoint introduced in Exercise 2.5.17.◇

3.8 Exercise [3]. Let Y and Z be V-valued random vectors with covariance operator C. Show that $\mathrm{Cov}(\phi(Y), \psi(Z)) = \phi(C(cv(\psi)))$ for all linear functionals ϕ and ψ on V. Deduce that even though C and $cv(\psi)$ (the coefficient vector of ψ) depend on the inner product employed on V, the linear transformation $\psi \to C(cv(\psi))$ does not.   ◇

3.9 Exercise [2]. Let V and W be inner product spaces. Show of V-valued random vectors $Y_1, \ldots, Y_m$ and W-valued random vectors $Z_1, \ldots, Z_n$ that

$$\mathrm{Cov}\left(\sum\nolimits_{1 \leq i \leq m} Y_i, \sum\nolimits_{1 \leq j \leq n} Z_j\right) = \sum\nolimits_{1 \leq i \leq m} \sum\nolimits_{1 \leq j \leq n} \mathrm{Cov}(Y_i, Z_n). \quad ◇$$

4. Dispersion operators

Background. If Y is an $\mathbb{R}^n$-valued random vector, it is customary to call the $n \times n$ matrix

$$\big(\mathrm{Cov}(Y_i, Y_j)\big)_{1 \le i \le n, 1 \le j \le n}$$

the variance-covariance matrix, or *dispersion matrix*, of Y. It is, of course, just the covariance matrix of Y and Y, in the sense of (3.1). ◇

In the coordinate-free context, the *dispersion operator* of Y taking values in V is defined to be $\mathrm{Cov}(Y, Y)$, denoted $\Sigma(Y)$, or Σ_Y, or simply Σ. From (3.5) we have

$$\Sigma = \Sigma', \text{ that is, } \Sigma \text{ is self-adjoint,} \tag{4.1}$$

while from (3.3) we have

$$\mathrm{Var}\big(\langle x, Y \rangle\big) = \mathrm{Cov}\big(\langle x, Y \rangle, \langle x, Y \rangle\big) = \langle x, \Sigma x \rangle \quad \text{for all } x \in V. \tag{4.2}$$

Thus Σ determines the variance of every linear functional of Y. By Exercise 2.5.13, the linear transformation Σ is uniquely determined by (4.1) and (4.2).

4.3 Exercise [3]. (1) Show that Y *has* a dispersion operator, in the sense that $\mathrm{Var}(\langle x, Y \rangle)$ is defined and finite for all $x \in V$, if and only if $E(\|Y\|^2) < \infty$.

(2) Show that a sufficient condition for V-valued random vectors Y and Z to *have* a covariance operator, in the sense that the left-hand side of (3.3) is defined and finite for all $y, z \in V$, is that $E(\|Y\|^2) < \infty$ and $E(\|Z\|^2) < \infty$. Is this condition necessary?

[Hint: For (1) show that $E(\|Y\|^2) < \infty$ implies $E(\langle x, Y \rangle^2) < \infty$ for all $x \in V$. For (2) show that $E(\|Y\|^2) < \infty$ and $E(\|Z\|^2) < \infty$ imply $E(|\langle y, Y \rangle|) < \infty$, $E(|\langle z, Z \rangle|) < \infty$, and $E(|\langle y, Y \rangle \langle z, Z \rangle|) < \infty$ for all $y, z \in V$.] ◇

4.4 Exercise [1]. Show that if $T: V \to W$ is linear, then

$$\Sigma_{TY} = T \Sigma_Y T', \tag{4.5}$$

the adjoint T' of T being taken in the sense of Exercise 2.5.17. ◇

4.6 Exercise [3]. Let $p_1, \ldots, p_k$ be strictly positive numbers summing to one and let $X = (X_1, \ldots, X_k)^T$ be a random vector in $\mathbb{R}^k$ taking the value $e^{(j)}$ (see (1.2)) with probability p_j for $j = 1, \ldots, k$. Define the random vector $Y = (Y_1, \ldots, Y_k)^T$ by

$$Y_j = \frac{X_j - p_j}{\sqrt{p_j}}, \quad j = 1, \ldots, k.$$

Show that Y has expectation 0 and dispersion operator $Q_{[r]} = P_{[r]^\perp}$, where r is the vector in $\mathbb{R}^k$ with j^{th} coordinate $\sqrt{p_j}$, $j = 1, \ldots, k$. ◇

As with covariance operators, dispersion operators depend on the inner product: By (3.4)

$$\Sigma_\Delta(Y) = (\Sigma(Y))\Delta, \tag{4.7}$$

where $\Sigma_\Delta(Y)$ denotes the dispersion operator of Y with respect to the $\langle \cdot, \cdot \rangle_\Delta$ inner product.

Y is said to have a *nonsingular distribution* if

$$\text{Var}(\langle x, Y \rangle) > 0 \quad \text{for all } 0 \neq x \in V, \tag{4.8}$$

that is, if all nonzero linear functionals of Y have strictly positive variance; otherwise Y is said to have a *singular distribution*. Letting

$$\Sigma = \sum_i \lambda_i P_{E_i}$$

be the spectral representation of Σ, so the E_i's are the mutually orthogonal eigenspaces of Σ and the λ_i's the corresponding eigenvalues, one sees that

$$\begin{aligned}
\text{Var}(\langle x, Y \rangle) = \langle x, \Sigma x \rangle &= \left\langle \sum_i P_{E_i} x, \sum_j \lambda_j P_{E_j} x \right\rangle \\
&= \sum_{ij} \lambda_j \langle P_{E_i} x, P_{E_j} x \rangle = \sum_i \lambda_i \| P_{E_i} x \|^2,
\end{aligned} \tag{4.9}$$

whence

$$Y \text{ has a nonsingular distribution}$$
$$\Longleftrightarrow \lambda_i > 0 \text{ for all } i \Longleftrightarrow \Sigma \text{ is nonsingular,}$$

whereas

$$Y \text{ has a singular distribution}$$
$$\Longleftrightarrow \lambda_i = 0 \text{ for some } i \Longleftrightarrow \Sigma \text{ is singular.}$$

Moreover, in the singular case,

$$\text{Var}(\langle x, Y \rangle) = 0 \quad \text{for all } x \in \mathcal{N}(\Sigma),$$

where $\mathcal{N}(\Sigma)$ denotes the null space of Σ, here the eigenspace of $\lambda = 0$, whereas

$$\text{Var}(\langle x, Y \rangle) > 0 \quad \text{for all nonzero } x \in (\mathcal{N}(\Sigma))^\perp = \mathcal{R}(\Sigma),$$

whence $Y - E(Y)$ takes all its values in $\mathcal{R}(\Sigma)$ with probability one and has in $\mathcal{R}(\Sigma)$ a nonsingular distribution.

4.10 Exercise [3]. Let Y be a V-valued random vector with mean μ and dispersion operator Σ. Show that $\mathcal{R}(\Sigma)$ is the smallest subspace M of V such that $Y - \mu$ takes values in M with probability one. ◇

4.11 Exercise [1]. Identify $\mathcal{R}(\Sigma)$ for the vector Y of Exercise 4.6. ◇

4.12 Exercise [2]. Suppose $n \geq 2$ and Y is a random vector in $\mathbb{R}^n$ with

$$\mathrm{Cov}(Y_i, Y_j) = \begin{cases} a, & \text{if } i = j, \\ b, & \text{if } i \neq j. \end{cases}$$

Identify $\mathcal{R}(\Sigma(Y))$ for those values of a and b such that Y has a singular distribution. [Hint: See Exercise 2.4.9.] ◇

4.13 Exercise [3]. Show that the nonsingularity of distributions is preserved by nonsingular linear transformations: If Y is nonsingular and $T: V \to V$ is nonsingular, then TY is nonsingular. Is it possible for TY to be nonsingular even though Y itself is singular? ◇

5. Weak sphericity

Motivation. If $Y_1, \ldots, Y_n$ are uncorrelated random variables each having variance σ^2, then the dispersion matrix of $Y = (Y_1, \ldots, Y_n)^T$ is $\sigma^2 I^{n \times n}$. In this case the variability

$$\mathrm{Var}(\langle c, Y \rangle) = \mathrm{Var}\left(\sum_i c_i Y_i\right) = \sigma^2 \sum_i c_i^2 = \sigma^2 \|c\|^2$$

of a linear combination of the Y_i's depends only on the length of the coefficient vector c and not on its orientation. ◇

In the coordinate-free setting, one says that a random vector Y taking values in V is *weakly spherical* (or has a *weakly spherical distribution*) if the dispersion operator of Y is proportional to the identity transformation $I: V \to V$ or, equivalently, if for some $\sigma^2 \geq 0$

$$\mathrm{Var}(\langle x, Y \rangle) = \sigma^2 \|x\|^2 \quad \text{for all } x \in V. \tag{5.1}$$

This says that the variability of Y, as measured by variance, is the same in all directions. The quantity σ^2 in (5.1) is called the *variance parameter* of Y. Y is said to have $\dim(V)$ *degrees of freedom*.

5.2 Exercise [3]. Show that Y is weakly spherical in V if and only if, for some $v \in V$, the random vectors $Y - v$ and $\mathcal{O}(Y - v)$ have the same mean and dispersion for all orthogonal linear transformations $\mathcal{O}: V \to V$. ◇

5.3 Exercise [2]. Show that if Y is weakly spherical in V, then Y is either nonsingular or constant with probability one. ◇

5.4 Exercise [3]. Let Y be weakly spherical in V with variance parameter σ^2 and let M be a subspace of V. Put $X = P_M Y$. Show that: (1) when X is considered as a V-valued random vector, it has the dispersion operator $\sigma^2 P_M$; (2) when X is considered as an M-valued random vector, it has the dispersion operator $\sigma^2 I_M$. ◇

A real random variable Y is automatically weakly spherical in $\mathbb{R}^1$, and for such a Y one has the standard formula

$$E(|Y|^2) = \mathrm{Var}(Y) + |E(Y)|^2$$

for the expected squared length of Y. To see how this formula generalizes to the arbitrary weakly spherical random vector, suppose Y is weakly spherical in V. Let $b_1, \ldots, b_n$ be an orthonormal basis for V. The $\langle b_i, Y \rangle$'s all have the same variance, say σ^2, and

$$E(\|Y\|^2) = E\Big(\sum_i \langle b_i, Y \rangle^2\Big) = \sum_i E\big(\langle b_i, Y \rangle^2\big)$$
$$= \sum_i \big[\mathrm{Var}\big(\langle b_i, Y \rangle\big) + \big(E\langle b_i, Y \rangle\big)^2\big]$$
$$= n\sigma^2 + \sum_i \langle b_i, EY \rangle^2 = n\sigma^2 + \|EY\|^2;$$

thus

$$E(\|Y\|^2) = \sigma^2 \dim(V) + \|EY\|^2. \tag{5.5}$$

6. Getting to weak sphericity

The distribution of a nonsingular random vector Y can easily be made weakly spherical by changing the inner product. Indeed, put $\Delta = c\Sigma^{-1}$, where $c > 0$ and Σ is the dispersion operator of Y. Then, by (4.7), the dispersion operator of Y in the $\langle \cdot, \cdot \rangle_\Delta$ inner product is

$$\Sigma\Delta = \Sigma(c\Sigma^{-1}) = cI,$$

so Y is weakly spherical with respect to $\langle \cdot, \cdot \rangle_\Delta$. As this argument shows, it is enough to know Σ up to a multiplicative factor.

6.1 Example. Suppose $Y_1, \ldots, Y_n$ are uncorrelated random variables with variances σ_i^2, $1 \le i \le n$. The matrix representing the dispersion operator Σ of $\boldsymbol{Y} = (Y_1, \ldots, Y_n)^T \in \mathbb{R}^n$ is diagonal with entries σ_i^2, $1 \le i \le n$, and for a given c the matrix representing $\Delta = c\Sigma^{-1}$ is diagonal with entries $w_i = c/\sigma_i^2$, $1 \le i \le n$. The above argument says that $\boldsymbol{Y}$ is weakly spherical with respect to the weighted dot-product

$$\langle \boldsymbol{x}, \boldsymbol{y} \rangle_w \equiv \langle \Delta\boldsymbol{x}, \boldsymbol{y} \rangle = \sum_i w_i x_i y_i; \tag{6.2}$$

note that the weights are proportional to reciprocal variances.   ●

6.3 Exercise [2]. Rewrite Example 2.1.17 for the case where the inner product is given by (6.2). Give explicit expressions for $P_{[e]}\boldsymbol{x}$ and $\|Q_{[e]}x\|^2$.   ◇

6.4 Exercise [2]. Supposing Y to be nonsingular (but not necessarily weakly spherical) in V, show that there exists a linear transformation $T: V \to V$ such that TY is weakly spherical (with respect to the original inner product). Exhibit such a T for the situation discussed in Example 6.1 above.
[Hint: In the general case, use (4.5) and Exercise 2.4.12.]   ◇

7. Normality

Motivation. If $Y_1, \ldots, Y_n$ are random variables having a joint normal distribution, then every linear combination $\langle c, \boldsymbol{Y} \rangle = \sum_i c_i Y_i$ of the Y_i's has a univariate normal distribution, and conversely. We make this the definition in the coordinate-free case.   ◇

A random vector Y taking values in V is said to have a *normal distribution* if

$$\langle v, Y \rangle \text{ is univariate normal for all } v \in V, \tag{7.1}$$

that is, if every linear functional of Y has a univariate normal distribution. Suppose Y is normal. Then $\langle v, Y \rangle$ has a finite mean and variance for all $v \in V$, so Y has an expectation $\mu \in V$ and a dispersion operator $\Sigma: V \to V$. From the fact that

$$\langle v, Y \rangle \sim N\big(\langle \mu, v \rangle, \langle v, \Sigma v \rangle\big)$$

(by definition of μ and Σ), it follows that Y has the characteristic function

$$\phi_Y(v) \equiv E(e^{i\langle v, Y \rangle}) = e^{i\langle \mu, v \rangle - \frac{1}{2}\langle v, \Sigma v \rangle}$$

($v \in V$). Just as a distribution on $\mathbb{R}^n$ is uniquely determined by its characteristic function, so too is a distribution on V determined by its characteristic function. It follows that μ and Σ determine the distribution of Y; we write

$$Y \sim N_V(\mu, \Sigma). \tag{7.2}$$

A linear transformation of a normal random vector is itself normal since for linear $T: V \to W$, $\langle w, TY \rangle_W = \langle T'w, Y \rangle_V$ for each $w \in W$. In fact, the formulas for propagating means and dispersions (see (2.4) and (4.5)) tell us that

$$Y \sim N_V(\mu, \Sigma) \quad \text{implies} \quad TY \sim N_W(T\mu, T\Sigma T'). \tag{7.3}$$

7.4 Exercise [1]. Show that a weakly spherical normal random vector Y has a (strictly) *spherical distribution,* in the sense that for all orthogonal linear transformations $\mathcal{O}: V \to V$, the random vectors $Y - \mu$ and $\mathcal{O}(Y - \mu)$ have the same distribution. ◇

Recall that if $\boldsymbol{X} = (X_1, \ldots, X_n)^T$ has a nonsingular normal distribution in $\mathbb{R}^n$ with $E(X_i) = \nu_i$ and $\mathrm{Cov}(X_i, X_j) = \sigma_{ij}$ for $1 \le i, j \le n$, then $\boldsymbol{X}$ has density

$$\frac{1}{(2\pi)^{n/2} |(\sigma_{ij})|^{1/2}} \, e^{-\frac{1}{2}\sum_{ij}(x_i - \nu_i)\sigma^{ij}(x_j - \nu_j)}, \quad (x_1, \ldots, x_n)^T \in \mathbb{R}^n, \tag{7.5}$$

with respect to Lebesgue measure on $\mathbb{R}^n$; here $|(\sigma_{ij})|$ and (σ^{ij}) are, respectively, the determinant and inverse of the matrix $(\sigma_{ij})_{1 \le i,j \le n}$. On a few occasions we will need the corresponding result for $Y \sim N_V(\mu, \Sigma)$ with Σ nonsingular, namely, that Y has density

$$\frac{1}{(2\pi)^{\dim(V)/2} |\Sigma|^{1/2}} \, e^{-\frac{1}{2}\langle y - \mu, \Sigma^{-1}(y - \mu)\rangle}, \quad y \in V, \tag{7.6}$$

with respect to *Lebesgue measure* λ on V; here λ is defined to be the image of Lebesgue measure on $\mathbb{R}^n$ under any isometric isomorphism (that is, distance-preserving linear transformation) from $\mathbb{R}^n$ to V, and $|\Sigma|$ denotes the determinant of the matrix of Σ with respect to any basis for V.

7.7 Exercise [5]. Check that λ above does not depend on the choice of the isometric isomorphism and that $|\Sigma|$ does not depend on the choice of basis, and deduce (7.6) from (7.5).

[Hint: Here is a more precise definition of λ. Let $S\colon \mathbb{R}^n \to V$ be an isometric isomorphism. For each Borel set B of V, $\lambda(B)$ is taken to be $\mu(S^{-1}(B))$, where μ is Lebesgue measure on $\mathbb{R}^n$. To work the exercise, use the following two facts from measure theory: (1) The image of μ under any orthogonal linear transformation of $\mathbb{R}^n$ to itself is μ itself. (2) If X_1, $\ldots$, X_n are random variables such that $\boldsymbol{X} = (X_1, \ldots, X_n)^T$ has a density f with respect to μ on $\mathbb{R}^n$ and if $S\colon \mathbb{R}^n \to V$ is an isometric isomorphism, then $S\boldsymbol{X}$ has the density fS^{-1} with respect to λ on V.]　　　　　　　　　　　　　　　　　　　　　　　　$\diamond$

8. The main result

We are now ready for the main result. The theoretical analysis of the GLM makes use of the resolution of a (weakly) spherical normal vector into several orthogonal components (subvectors), which by Theorem 8.2 below are independent and have themselves spherical normal distributions, albeit in spaces of lower dimensionality.

Recall that the $\chi^2_{\nu;\theta}$ distribution (χ^2 *with ν degrees of freedom and noncentrality parameter $\theta \geq 0$*) is defined to be the distribution of

$$Z_1^2 + \cdots + Z_\nu^2,$$

with the Z_i's being independent normal random variables of unit variance and with expectations $EZ_1 = \theta$, $EZ_2 = \cdots = EZ_\nu = 0$. The distribution is said to be *central* if $\theta = 0$ and *noncentral* if $\theta > 0$. In the central case one often writes χ^2_ν in place of $\chi^2_{\nu;0}$.

8.1 Exercise [3]. Suppose S is a random variable distributed according to the $\chi^2_{\nu;\theta}$ distribution. Show that

$$E(S) = \nu + \theta^2 \quad \text{and} \quad \text{Var}(S) = 2\nu + 4\theta^2. \qquad\qquad \diamond$$

8.2 Theorem. *Let Y be a random vector in V having a weakly spherical distribution with variance parameter σ^2:*

$$\Sigma_Y = \sigma^2 I.$$

Let $M_1, \ldots, M_k$ be mutually orthogonal subspaces of V. Then

(i) *the random vectors $P_{M_i} Y$ for $i = 1, \ldots, k$ are uncorrelated, have weakly spherical distributions in their respective subspaces, and*

$$E(P_{M_i} Y) = P_{M_i}(EY), \quad E\|P_{M_i} Y\|^2 = \sigma^2 \dim(M_i) + \|P_{M_i}(EY)\|^2. \quad (8.3)$$

If, in addition, Y is normally distributed, then

(ii) *the random vectors $P_{M_i} Y$ are independent and normally distributed, and the random variables $\|P_{M_i} Y\|^2/\sigma^2$ are independent with χ^2 distributions of $\dim(M_i)$ degrees of freedom and noncentrality parameters $\|P_{M_i}(EY)\|/\sigma$:*

$$\|P_{M_i} Y\|^2 \sim \sigma^2 \chi^2_{\dim(M_i);\|P_{M_i}(EY)\|/\sigma} \quad \text{for } 1 \leq i \leq k. \qquad (8.4)$$

Proof. (i) *holds*: The $P_{M_i}Y$'s are uncorrelated since, by (3.6),

$$\text{Cov}(P_{M_i}Y, P_{M_j}Y) = \sigma^2 P_{M_i} I P_{M_j} = \sigma^2 P_{M_i} P_{M_j} = 0$$

for $i \neq j$. $P_{M_i}Y$ has a weakly spherical distribution in M_i since

$$\text{Var}(\langle m, P_{M_i}Y\rangle) = \text{Var}(\langle P_{M_i}m, Y\rangle) = \text{Var}(\langle m, Y\rangle) = \sigma^2\|m\|^2$$

for all $m \in M_i$. The formulas for the expectations of $P_{M_i}Y$ and $\|P_{M_i}Y\|^2$ are restatements of earlier results ((2.4) and (5.5)).

(ii) *holds*: Assume for notational convenience that $V = M_1 + M_2$ with $M_1 \perp M_2$. Let $b_1, \ldots, b_d$ and $b_{d+1}, \ldots, b_n$ be orthonormal bases for M_1 and M_2, respectively ($d = \dim(M_1)$, $n = \dim(V)$). Without loss of generality, let b_1 point in the direction of $P_{M_1}(EY)$ and let b_{d+1} point in the direction of $P_{M_2}(EY)$ when these components of EY are nonzero. Then

$$P_{M_1}Y = \sum_{1 \leq i \leq d}\langle b_i, Y\rangle b_i \quad\text{and}\quad P_{M_2}Y = \sum_{d+1 \leq i \leq n}\langle b_i, Y\rangle b_i$$

and

$$\|P_{M_1}Y\|^2 = \sum_{1 \leq i \leq d}\langle b_i, Y\rangle^2 \quad\text{and}\quad \|P_{M_2}Y\|^2 = \sum_{d+1 \leq i \leq n}\langle b_i, Y\rangle^2.$$

The assertions of (ii) follow from the fact that the uncorrelated random variables $\langle b_1, Y\rangle$, ..., $\langle b_n, Y\rangle$ are independently normally distributed with common variance σ^2; $E\langle b_i, Y\rangle = 0$ for $i \neq 1, d+1$, while

$$E\langle b_1, Y\rangle = E\langle\frac{P_{M_1}(EY)}{\|P_{M_1}(EY)\|}, Y\rangle = \frac{\langle P_{M_1}(EY), EY\rangle}{\|P_{M_1}(EY)\|} = \|P_{M_1}(EY)\|$$

($= 0$, if $P_{M_1}(EY) = 0$) and $E\langle b_{d+1}, Y\rangle = \|P_{M_2}(EY)\|$. ∎

8.5 Exercise [2]. Show that: (1) if $X_1, \ldots, X_\nu$ are independent normal random variables with unit variances, then $\sum_{i=1}^{\nu} X_i^2 \sim \chi^2_{\nu;\theta}$ with $\theta = \sqrt{\sum_i (EX_i)^2}$; (2) if S_1 and S_2 are independent random variables with $S_i \sim \chi^2_{\nu_i;\theta_i}$ for $i = 1, 2$, then $S_1 + S_2 \sim \chi^2_{\nu_1+\nu_2;\sqrt{\theta_1^2+\theta_2^2}}$. ◇

8.6 Exercise [4]. Let $\boldsymbol{X}^{(1)}, \ldots, \boldsymbol{X}^{(n)}$ be independent random vectors in $\mathbb{R}^k$, each having the distribution of the random vector $\boldsymbol{X}$ in Exercise 4.6. Put

$$\boldsymbol{N} = \sum_{1 \leq m \leq n} \boldsymbol{X}^{(m)};$$

the distribution of $\boldsymbol{N} = (N_1, \ldots, N_k)^T$ is the multinomial distribution for n trials and k cells having occupancy probabilities $p_1, \ldots, p_k$. Show that the limiting distribution, taken as $n \to \infty$, of Pearson's *goodness of fit statistic*

$$G_n \equiv \sum_{1 \leq j \leq k} \frac{(N_j - np_j)^2}{np_j}$$

is χ^2_{k-1}.

[Hint: Observe that G_n can be written as $\|T_n\|^2$, where

$$T_n = \frac{1}{\sqrt{n}} \sum_{1 \leq m \leq n} Y^{(m)}$$

with

$$Y_j^{(m)} = \frac{X_j^{(m)} - p_j}{\sqrt{p_j}}$$

for $j = 1, \ldots, k$. Deduce from Exercises 4.6 and 5.4 and the multivariate central limit theorem that the limiting distribution of G_n is that of $\|Q_{[\mathbf{r}]}Z\|^2$, where $Z \sim N_{\mathbb{R}^k}(0, I_{\mathbb{R}^k})$ and $\mathbf{r}$ is the vector specified in Exercise 4.6.] $\diamond$

We will have several occasions to make use of the following representation of the noncentral χ^2 distribution as a Poisson mixture of central χ^2 distributions.

8.7 Proposition. *For any $\nu \geq 1$ and $\theta \geq 0$, one has*

$$\chi^2_{\nu;\theta} = \sum_{0 \leq k < \infty} e^{-\lambda} \frac{\lambda^k}{k!} \chi^2_{\nu+2k}, \qquad (8.8)$$

where

$$\lambda = \theta^2/2. \qquad (8.9)$$

Proof. Let $Z_1, \ldots, Z_\nu$ be independent random variables with $Z_i \sim N(\theta_i, 1)$ for $1 \leq i \leq \nu$ and with

$$\theta_1 = \theta, \quad \theta_2 = \theta_3 = \cdots = \theta_\nu = 0.$$

Put

$$X = Z_1^2 + \cdots + Z_\nu^2,$$

so that X has distribution $\chi^2_{\nu;\theta}$. The moment generating function of X is

$$E(e^{tX}) = \prod_{1 \leq i \leq \nu} E(e^{tZ_i^2}).$$

For $t < \frac{1}{2}$, the technique of "completing the square" gives

$$E(e^{tZ_i^2}) = \int_{-\infty}^{\infty} e^{tz^2} \frac{1}{\sqrt{2\pi}} e^{-\frac{1}{2}(z-\theta_i)^2} dz$$

$$= e^{-\frac{1}{2}\theta_i^2(1-\frac{1}{1-2t})} \int_{-\infty}^{\infty} \frac{1}{\sqrt{2\pi}} e^{-\frac{1}{2}\left((1-2t)z^2 - 2z\theta_i + \frac{\theta_i^2}{1-2t}\right)} dz$$

$$= e^{-\frac{1}{2}\theta_i^2(1-\frac{1}{1-2t})} \frac{1}{\sqrt{1-2t}},$$

whence

$$E(e^{tX}) = e^{-\lambda}\, e^{\lambda \frac{1}{1-2t}} \Big(\frac{1}{1-2t} \Big)^{\nu/2} = \sum\nolimits_{0 \le k < \infty} e^{-\lambda} \frac{\lambda^k}{k!} \Big(\frac{1}{1-2t} \Big)^{(\nu+2k)/2}. \qquad (8.10)$$

Setting θ, and so also λ, equal to 0 here shows that the χ^2_ν distribution has the moment generating function $(1-2t)^{-\nu/2}$. It follows that the right-hand side of (8.10) is the moment generating function of the right-hand side of (8.8), and an appeal to the uniqueness theorem for moment generating functions finishes the proof. ∎

8.11 Exercise [2]. Suppose S is a random variable distributed according to the $\chi^2_{\nu;\theta}$ distribution. Show that

$$E\Big(\frac{1}{S}\Big) = \begin{cases} \sum\nolimits_{0 \le k < \infty} e^{-\lambda} \dfrac{\lambda^k}{k!} \dfrac{1}{\nu - 2 + 2k}, & \text{if } \nu \ge 3; \\ \infty, & \text{if } \nu \le 2. \end{cases} \qquad (8.12)$$

[Hint: Recall that the χ^2_ν distribution has density $x^{\frac{1}{2}\nu - 1} e^{-x/2} / (2^{\frac{1}{2}\nu} \Gamma(\frac{1}{2}\nu))$ for $x > 0$.] ◇

8.13 Exercise [3]. The $\mathcal{F}^*_{\nu_1,\nu_2;\theta}$ distribution (*unnormalized $\mathcal{F}$ with ν_1 and ν_2 degrees of freedom and noncentrality parameter θ*) is defined to be the distribution of

$$\frac{S_1}{S_2}, \qquad (8.14)$$

where S_1 and S_2 are independent random variables with $S_1 \sim \chi^2_{\nu_1;\theta}$ and $S_2 \sim \chi^2_{\nu_2}$; $\mathcal{F}^*_{\nu_1,\nu_2}$ is short for $\mathcal{F}^*_{\nu_1,\nu_2;0}$. Show that for any $\nu_1, \nu_2 \ge 1$ and $\theta \ge 0$,

$$\mathcal{F}^*_{\nu_1,\nu_2;\theta} = \sum\nolimits_{0 \le k < \infty} e^{-\lambda} \frac{\lambda^k}{k!} \mathcal{F}^*_{\nu_1+2k,\nu_2}, \qquad (8.15)$$

with $\lambda = \theta^2/2$. ◇

9. Problem set: Distribution of quadratic forms

A *quadratic form* in n (real) variates $x_1, \ldots, x_n$ is an expression of the form

$$Q(\boldsymbol{x}) = \sum\nolimits_{ij} x_i a_{ij} x_j = \langle \boldsymbol{x}, \boldsymbol{A}\boldsymbol{x} \rangle = \boldsymbol{x}^T \boldsymbol{A} \boldsymbol{x} = \boldsymbol{x}^T \boldsymbol{A}^T \boldsymbol{x} = \boldsymbol{x}^T \Big(\frac{\boldsymbol{A} + \boldsymbol{A}^T}{2} \Big) \boldsymbol{x}.$$

There being no loss of generality, one customarily takes the matrix $\boldsymbol{A}$ to be symmetric.

To put this idea into the coordinate-free context, we define a *quadratic form* on V to be a mapping $Q \colon V \to \mathbb{R}$ of the form

$$Q(v) = \langle v, Tv \rangle,$$

where $T \colon V \to V$ is linear and self-adjoint. By Exercise 2.5.13, T is uniquely determined by Q. This problem set develops some properties of $Q(Y)$ where Y is a V-valued random vector. Throughout Q, with or without subscripts, denotes a quadratic form on V with associated self-adjoint operator T. The properties of orthogonal projections established in Section 2.2 should be employed wherever appropriate.

A. Suppose Y is weakly spherical with variance parameter σ^2. Show that

$$E\big(Q(Y)\big) = \sigma^2 \operatorname{trace}(T) + Q(EY), \qquad (9.1)$$

where trace(T) is the *trace* of T, that is, the sum of the eigenvalues of T, multiplicities included; relate this to (5.5).
[Hint: Use the spectral decomposition of T.]   o

For parts **B** through **G**, suppose $Y \sim N_V(\mu, I)$ for some $\mu \in V$. Show:

B. One has

$$\operatorname{Var}\big(Q(Y)\big) = 2 \operatorname{trace}(T^2) + 4\|T(\mu)\|^2. \qquad (9.2)\ \text{o}$$

C. $Q(Y)$ has a χ^2 distribution $\Longleftrightarrow T$ is idempotent (that is, a projection), in which case

$$Q(Y) \sim \chi^2_{\nu;\theta} \text{ with } \nu = \rho(T) \equiv \dim\big(\mathcal{R}(T)\big) \text{ and } \theta = \sqrt{Q(\mu)}. \qquad (9.3)$$

[Hint: For ($\Longrightarrow$) use the spectral decomposition of T, moment generating functions, and Proposition 8.7.]   o

D. Suppose that $Q(v) = Q_1(v) + Q_2(v)$ for all $v \in V$ and that $Q(Y) \sim \chi^2_{\nu;\theta}$ and $Q_1(Y) \sim \chi^2_{\nu_1;\theta_1}$. Then $Q_2(Y)$ has a χ^2 distribution $\Longleftrightarrow Q_2(v) \geq 0$ for all v, in which case

$$Q_2(Y) \sim \chi^2_{\nu-\nu_1;\sqrt{\theta^2-\theta_1^2}} \text{ and is independent of } Q_1(Y).$$

[Hint: Use part **C** and Proposition 2.2.13 applied to $T_2 = T - T_1$.]   o

E. Suppose $Q(v) = Q_1(v) + Q_2(v)$ for all $v \in V$ and that $Q_i(Y) \sim \chi^2_{\nu_i;\theta_i}$ for $i = 1, 2$. Then $Q(Y)$ has a χ^2 distribution $\Longleftrightarrow Q_1(Y)$ and $Q_2(Y)$ are independent $\Longleftrightarrow T_1 T_2 = 0$, in which case

$$Q(Y) \sim \chi^2_{\nu_1+\nu_2;\sqrt{\theta_1^2+\theta_2^2}}. \qquad (9.4)$$

[Hint: See Proposition 2.2.24.]   o

F. (Cochran's theorem).
Suppose for given k and $Q_1, \ldots, Q_k$ that $Q(v) = \sum_{1 \leq i \leq k} Q_i(v)$ for all $v \in V$ and that $Q(Y)$ has a χ^2 distribution. Then the $Q_i(Y)$'s have χ^2 distributions $\Longleftrightarrow$ one of the following equivalent conditions hold:

   (i) the T_i's are idempotent,
   (ii) $T_i T_j = 0$ for $i \neq j$,
   (iii) $\sum_i \rho(T_i) = \rho(T)$,

and then

$$Q_i(Y) \sim \chi^2_{\rho(T_i), \sqrt{Q_i(\mu)}} \text{ for each } i \text{ and the } Q_i(Y)\text{'s are independent.} \qquad (9.5) \circ$$

G. Formulate and prove a distribution analog of Exercise 2.2.18. $\circ$

H. Generalize parts **B** through **F** to the case

$$Y \sim N_V(\mu, \Sigma),$$

where Σ is nonsingular.

[Hint: Change the inner product to get back to the spherical case.] $\circ$

I. Apply Cochran's theorem to the standard decomposition

$$\sum_i Y_i^2 = \sum_i (Y_i - \bar{Y})^2 + n\bar{Y}^2$$

of the sum of squares in the one-sample problem

$$Y \sim N(\mu e, \sigma^2 I).$$

[Hint: See Exercise 2.2.42.] $\circ$

CHAPTER 4

GAUSS-MARKOV ESTIMATION

Throughout this chapter we suppose that

> the random vector Y is weakly spherical $\big(\Sigma(Y) = \sigma^2 I\big)$
> in a given inner product space $(V, \langle \cdot, \cdot \rangle)$ and has an un-
> known mean μ lying in some given subspace M of V. $\qquad(0.1)$

M is often called the *regression manifold*. The variance parameter σ^2 may be known or unknown; we assume it to be greater than 0 to avoid some trivial considerations. We will develop point estimates of μ (and of σ^2, in the case where σ^2 is unknown) and exhibit some of the properties of these estimators.

Now, to know μ is to know the values of all linear functionals of μ, and vice versa. So it is in a sense immaterial whether one estimates μ itself or the linear functionals of μ. Since in many applications of the GLM certain linear functionals of μ are the items of prime interest, we will approach the business of estimating μ initially from the linear functional point of view.

1. Linear functionals of μ

We already know that for any given linear functional ψ on M, there exists a unique vector $cv(\psi)$ in M, called the *coefficient vector* of ψ, such that

$$\psi(m) = \langle cv(\psi), m \rangle \quad \text{for all } m \in M. \qquad(1.1)$$

Often the linear functional will be given initially in the form

$$\psi(m) = \langle x, m \rangle$$

$(m \in M)$ for some $x \in V$. Because $\langle x, m \rangle = \langle P_M x, m \rangle$ for all $m \in M$, we necessarily have

$$cv(\psi) = P_M x \qquad(1.2)$$

in this case. For ease of notation, it is convenient to define an *inner product* and *norm* for linear functionals on M as follows:

$$\langle \psi_1, \psi_2 \rangle = \langle cv(\psi_1), cv(\psi_2) \rangle, \tag{1.3}$$

$$\|\psi\| = \|cv(\psi)\|. \tag{1.4}$$

1.5 Example *(The triangle problem).* Suppose $V = \mathbb{R}^3$ and

$$M = \left\{ \sum_{1 \le j \le 3} \beta_j e^{(j)} : \beta_1 + \beta_2 + \beta_3 = 0 \right\}$$

$$= \{ x \in \mathbb{R}^3 : x_1 + x_2 + x_3 = 0 \} = [e]^\perp$$

$(e_i^{(j)} = \delta_{ij}$ and $e = e^{(1)} + e^{(2)} + e^{(3)})$. This regression manifold arises in connection with the problem of determining the three angles of a triangle, which must sum to $180°$; think of β_j as being the jth angle, less $60°$. Consider the linear functional ψ_j on M defined by

$$\psi_j \left(\sum_i \beta_i e^{(i)} \right) = \beta_j = \left\langle \sum_i \beta_i e^{(i)}, e^{(j)} \right\rangle.$$

$e^{(j)}$ is *not* the coefficient vector of ψ_j, because $e^{(j)} \notin M$. We may obtain $cv(\psi_j)$ as follows:

$$cv(\psi_j) = P_M e^{(j)} = e^{(j)} - P_{[e]} e^{(j)} \tag{1.6}$$

$$= e^{(j)} - \frac{\langle e^{(j)}, e \rangle}{\langle e, e \rangle} e = e^{(j)} - \frac{1}{3} e. \qquad \bullet$$

In many instances M is specified in terms of a basis; in such cases the associated coordinate functionals are of special interest. Suppose then that $x_1, \ldots, x_p$ is a basis for M:

$$M = \left\{ \sum_{1 \le i \le p} \beta_i x_i : \beta_1, \ldots, \beta_p \in \mathbb{R} \right\}. \tag{1.7}$$

Let ψ_j be the jth-coordinate functional on M with respect to this basis:

$$\psi_j \left(\sum_i \beta_i x_i \right) = \beta_j.$$

ψ_j is a linear functional on M. To obtain its coefficient vector, we set

$$M_j = [x_1, \ldots, x_{j-1}, \ x_{j+1}, \ldots, x_p],$$

write the generic $x \in M$ as

$$x = \sum_i \beta_i x_i = \beta_j Q_j x_j + \left(\sum_{i \ne j} \beta_i x_i + \beta_j P_j x_j \right)$$

with $P_j = P_{M_j}$ and $Q_j = Q_{M_j}$, and use orthogonality to get

$$\langle x, Q_j x_j \rangle = \beta_j \|Q_j x_j\|^2.$$

Since $0 \neq Q_j x_j = x_j - P_j x_j \in M$, we obtain

$$cv(\psi_j) = \left(\frac{Q_j x_j}{\|Q_j x_j\|}\right) \frac{1}{\|Q_j x_j\|} \quad \text{and} \quad \|\psi_j\| = \frac{1}{\|Q_j x_j\|}, \qquad (1.8)$$

as asserted in Exercise 2.5.3. Note that $Q_j x_j = x_j$ if x_j is perpendicular to each of $x_1, \ldots, x_{j-1}, x_{j+1}, \ldots, x_p$, in particular if the x_i's are an orthogonal basis for M.

1.9 Example. In simple linear regression one has $V = \mathbb{R}^n$ and

$$M = \{\alpha e + \beta x : \alpha, \beta \in \mathbb{R}\} = [e, x],$$

with x a given vector in $\mathbb{R}^n$ that is not a multiple of e. Let S be the *slope functional* on M:

$$S(\alpha e + \beta x) = \beta.$$

By (1.8), its coefficient vector is

$$cv(S) = \frac{Q_{[e]}(x)}{\|Q_{[e]}(x)\|^2} = \left(\frac{x_i - \bar{x}}{\sum_j (x_j - \bar{x})^2}\right)_{1 \leq i \leq n} \qquad (1.10)$$

because

$$Q_{[e]}(x) = x - P_{[e]}(x) = x - \frac{\langle x, e \rangle}{\langle e, e \rangle} e = x - \bar{x} e. \qquad \diamond$$

1.11 Exercise [2]. In the basis form (1.7) of the GLM, write

$$z_i \equiv cv(\psi_i) = \frac{Q_i x_i}{\|Q_i x_i\|^2} = \sum_{1 \leq j \leq p} c_{ij} x_j, \quad 1 \leq i \leq p. \qquad (1.12)$$

Show that

$$C \equiv (c_{ij}) = A^{-1}, \qquad (1.13)$$

where

$$A = (a_{jk}) \quad \text{with} \quad a_{jk} = \langle x_j, x_k \rangle. \qquad (1.14)$$

Show also that

$$c_{ik} = \langle z_i, z_k \rangle. \qquad (1.15)$$

[Hint: $\psi_i(x_k) = \delta_{ik}$.] $\diamond$

2. Estimation of linear functionals of μ

Since by assumption the variability of Y (as measured by variance) is the same in all directions and since $P_M Y$ is the closest point in M to Y, it is intuitively plausible that $P_M Y$ should be a decent estimator of μ. And, of course, if we agree to estimate μ by $P_M Y$, it is natural to estimate the linear functional $\psi(\mu)$ by $\psi(P_M Y)$.

The *Gauss-Markov estimator (GME)*, $\hat{\psi}(Y)$, of a linear functional $\psi(\mu)$ of μ is defined by

$$\hat{\psi} \equiv \hat{\psi}(Y) = \psi(P_M Y) = \langle cv(\psi), P_M Y \rangle = \langle cv(\psi), Y \rangle. \qquad (2.1)$$

Notice that for $x \in V$ the GME of the linear functional $\mu \to \langle x, \mu \rangle$ is

$$\langle P_M x, Y \rangle = \langle P_M x, P_M Y \rangle = \langle x, P_M Y \rangle;$$

one must project either Y, or x, or both onto M before taking the inner product. In particular, when $x \in M$, $\langle x, Y \rangle$ is the GME of its expected value $\langle x, \mu \rangle$; this observation can frequently be used to obtain GMEs more or less at sight. To put it another way, if for a given linear functional ψ on M we can (aided by statistical intuition) guess at an $x \in M$ such that $E_\mu \langle x, Y \rangle = \psi(\mu)$ for all $\mu \in M$, then $\hat{\psi}(Y) = \langle x, Y \rangle$.

2.2 Example. (A) In the triangle problem (Example 1.5), the GME of β_j is (see (1.6))

$$\hat{\beta}_j = \langle cv(\psi_j), Y \rangle = \langle e^{(j)} - \frac{1}{3}e, Y \rangle = Y_j - \frac{Y_1 + Y_2 + Y_3}{3}.$$

(B) In simple linear regression, the GME of the slope functional $S(\mu)$ is (see (1.10))

$$\hat{\beta} = \langle cv(S), Y \rangle = \langle \frac{x - \bar{x}e}{\|x - \bar{x}e\|^2}, Y \rangle = \frac{\sum_i (x_i - \bar{x})Y_i}{\sum_i (x_i - \bar{x})^2}.$$

(C) In the context of the basis case (1.7) of the GLM, $\langle x_j, Y \rangle$ is the GME of its expected value $\langle x_j, \sum_{1 \le k \le p} \beta_k x_p \rangle = \sum_{1 \le k \le p} a_{jk} \beta_k$, where $a_{jk} = \langle x_j, x_k \rangle$.

(D) As an example of "guessing," consider one-way analysis of variance where

$$V = \{ (x_{ij})_{j=1,\ldots,n_i; \, i=1,\ldots,p} \in \mathbb{R}^n \}$$

with $n = \sum_i n_i$ and

$$M = \{ \sum_{1 \le i \le p} v_i \beta_i : \beta_1, \ldots, \beta_p \in \mathbb{R} \} \tag{2.3}$$

with $(v_i)_{i'j'} = \delta_{ii'}$. β_i is the common expectation of $Y_{i1}, \ldots, Y_{in_i}$, so an obvious guess as to its GME is the sample mean $\bar{Y}_i = \sum_j Y_{ij}/n_i = \langle Y, v_i/n_i \rangle$. This hunch is correct, because $E(\bar{Y}_i) = \beta_i$ and $v_i/n_i \in M$. ●

2.4 Exercise [1]. Show that if $\psi_1(\mu), \ldots, \psi_k(\mu)$ are linear functionals of μ and $a_1, \ldots, a_k$ are constants, then the GME of $\sum_{1 \le i \le k} a_i \psi_i(\mu)$ is $\sum_{1 \le i \le k} a_i \hat{\psi}_i$. ◇

2.5 Exercise [1]. Consider the basis form (1.7) of the GLM. Use the assertions of Example 2.2(C) and the preceding exercise to show that the GMEs $\hat{\beta}_1, \ldots, \hat{\beta}_p$ of $\beta_1, \ldots, \beta_p$ satisfy the normal equations $\langle x_j, Y \rangle = \sum_{1 \le k \le p} \langle x_j, x_k \rangle \hat{\beta}_k$, $1 \le j \le k$. ◇

The GME $\hat{\psi}(Y)$ of a linear functional $\psi(\mu)$ has the following properties:

$$\hat{\psi}(Y) \text{ is a linear function of } Y, \tag{2.6}$$

$$\hat{\psi}(Y) \text{ is an } \textit{unbiased estimator} \text{ of } \psi(\mu), \text{ that is,}$$
$$E_\mu(\hat{\psi}(Y)) = \psi(\mu) \quad \text{for all } \mu \in M \tag{2.7}$$

(because $E_\mu \hat{\psi}(Y) = E_\mu \langle cv(\psi), Y \rangle = \langle cv(\psi), \mu \rangle = \psi(\mu)$), and

$$\mathrm{Var}(\hat{\psi}(Y)) = \sigma^2 \|cv(\psi)\|^2 = \sigma^2 \|\psi\|^2 \tag{2.8}$$

(see (1.4)). Moreover, $\hat{\psi}(Y)$ is the *best linear unbiased estimator (BLUE)* of $\psi(\mu)$ in the following sense.

2.9 Theorem (Gauss-Markov theorem). *For each linear functional ψ of μ, the GME $\hat{\psi}(Y)$ is the unique estimator having minimum variance in the class of linear unbiased estimators of $\psi(\mu)$.*

Proof. Suppose for a given $x \in V$, $\langle x, Y \rangle$ unbiasedly estimates $\psi(\mu)$, so that

$$\psi(\mu) = E_\mu(\langle x, Y \rangle) = \langle x, \mu \rangle$$

for every element μ of M. Then $cv(\psi) = P_M x$ by (1.2), whence

$$\mathrm{Var}(\langle x, Y \rangle) = \sigma^2 \|x\|^2 \geq \sigma^2 \|P_M x\|^2 = \sigma^2 \|cv(\psi)\|^2 = \mathrm{Var}(\hat{\psi}(Y))$$

with equality holding if and only if $x = P_M x$, that is, $\langle x, Y \rangle = \hat{\psi}(Y)$. ∎

Note that the covariance between two GMEs $\hat{\psi}_1$ and $\hat{\psi}_2$ is given by

$$\mathrm{Cov}(\hat{\psi}_1(Y), \hat{\psi}_2(Y)) = \sigma^2 \langle cv(\psi_1), cv(\psi_2) \rangle = \sigma^2 \langle \psi_1, \psi_2 \rangle; \tag{2.10}$$

in particular, $\hat{\psi}_1(Y)$ and $\hat{\psi}_2(Y)$ are uncorrelated if and only if $cv(\psi_1)$ and $cv(\psi_2)$ are orthogonal.

2.11 Example. (A) In the triangle problem (Example 1.5), the covariance between $\hat{\beta}_i$ and $\hat{\beta}_j$ is

$$\begin{aligned}
\mathrm{Cov}(\hat{\beta}_i, \hat{\beta}_j) &= \sigma^2 \langle Q_{[e]} e^{(i)}, Q_{[e]} e^{(j)} \rangle \\
&= \sigma^2 (\langle e^{(i)}, e^{(j)} \rangle - \langle P_{[e]} e^{(i)}, P_{[e]} e^{(j)} \rangle) \\
&= \sigma^2 (\delta_{ij} - \|e/3\|^2) = \sigma^2 (\delta_{ij} - 1/3).
\end{aligned}$$

(B) In the basis form (1.7) of the GLM, the variance of the GME of $\hat{\beta}_j$ is

$$\mathrm{Var}(\hat{\beta}_j) = \sigma^2 / \|Q_j x_j\|^2; \tag{2.12}$$

this is larger the more collinear x_j is with the other x_k's. As a special case of (2.12), we get

$$\mathrm{Var}(\hat{\beta}) = \frac{\sigma^2}{\|Q_{[e]} x\|^2} = \frac{\sigma^2}{\sum_j (x_j - \bar{x})^2}$$

for the variance of the slope coefficient $\hat{\beta}$ in simple linear regression. •

2.13 Exercise [1]. Suppose in the standard setup

$$E(Y) = X\beta, \quad \Sigma(Y) = \sigma^2 I$$

of the GLM in $\mathbb{R}^n$, one of the columns of the $n \times p$ full rank matrix $X = (x_{ij})$ consists solely of 1's. The functional

$$\psi(\mu) = \sum_{1 \le i \le p} \bar{x}_{\cdot i} \beta_i$$

$(\bar{x}_{\cdot i} = \frac{1}{n} \sum_{1 \le m \le n} x_{mi}$ for $1 \le i \le p)$ gives the expected "response" at the mean values of the independent variables. Find the GME of $\psi(\mu)$ and show that it has standard deviation simply $\sigma/\sqrt{n}$.
[Hint: A one-line calculation gives the coefficient vector of ψ.] $\diamond$

2.14 Exercise [4]. In the two-way additive analysis of variance layout with one observation per cell, one has $Y = (Y_{ij})_{1 \le i \le I, 1 \le j \le J}$ weakly spherical in the space V of $I \times J$ arrays endowed with the dot-product of Exercise 2.2.26 and

$$\mu_{ij} = E(Y_{ij}) = \nu + \alpha_i + \beta_j \quad \text{for } 1 \le i \le I, 1 \le j \le J, \tag{2.15_1}$$

for numbers $\nu, \alpha_1, \ldots, \alpha_I, \beta_1, \ldots, \beta_J$ satisfying

$$\sum_i \alpha_i = 0 = \sum_j \beta_j . \tag{2.15_2}$$

ν is called the *grand mean*, α_i the *differential effect for row* i, and β_j the *differential effect for column* j. Show that the set M of arrays $\mu = (\mu_{ij})$ satisfying (2.15) is a subspace of V and that the mappings $\mu \to \nu$, $\mu \to \alpha_i$, and $\mu \to \beta_j$ are linear functionals on M. Find the coefficient vectors of these linear functionals and show that the GMEs of ν, α_i, and β_j are, respectively,

$$\hat{\nu} = \bar{Y}_{\cdot\cdot}, \quad \hat{\alpha}_i = \bar{Y}_{i\cdot} - \bar{Y}_{\cdot\cdot}, \quad \text{and} \quad \hat{\beta}_j = \bar{Y}_{\cdot j} - \bar{Y}_{\cdot\cdot}, \tag{2.16}$$

where

$$\bar{Y}_{\cdot\cdot} = \frac{1}{IJ} \sum_{ij} Y_{ij}, \quad \bar{Y}_{i\cdot} = \frac{1}{J} \sum_j Y_{ij}, \quad \text{and} \quad \bar{Y}_{\cdot j} = \frac{1}{I} \sum_i Y_{ij} .$$

Use (2.10) to compute the variances and covariances of $\hat{\nu}, \hat{\alpha}_1, \ldots, \hat{\alpha}_I$, and $\hat{\beta}_1, \ldots, \hat{\beta}_J$.
[Hint: $cv(\alpha_i) = \frac{1}{J} Q_{[e]} r_i$, where r_i and e are the $I \times J$ matrices with $(r_i)_{i'j'} = \delta_{ii'}$ and $e_{ij} = 1$.] $\diamond$

2.17 Exercise (*The four-penny problem*) [4]. Suppose you have four pennies, say P_1, P_2, P_3, and P_4, a weighing balance with two pans, and a set of weights. You are allowed to make only four weighings of the pennies. In each weighing, you may place some (possibly none) of the pennies in the left pan, some (possibly none) in the right pan, and leave the remaining pennies (if any) off the balance. The problem is this: How should the four weighings be designed so as to produce the most accurate determination of the weights of the four pennies, in the sense that the sum of the variances of the four weight estimators should be as small as possible? Assume that weighings are independent and that measuring errors have mean 0 and a common variance, say σ^2, that does not depend on the configuration of pennies on the scale.

To put this problem in the context of the GLM, let β_1, β_2, β_3, and β_4 be the unknown weights of the four pennies. For $i = 1, \ldots, 4$, let W_i be the random variable recording the result of the ith weighing; adopt the convention that weights placed in the left (respectively, right) pan are treated as being positive (respectively, negative). The assumptions of the problem imply that for any given weighing design, the random vector $\boldsymbol{W} = (W_1, W_2, W_3, W_4)^T$ is weakly spherical and for each i,

$$E(W_i) = \sum_{1 \leq j \leq 4} x_{ij} \beta_j,$$

where

$$x_{ij} = \begin{cases} -1, & \text{if penny } j \text{ is to be placed in the left pan} \\ 0, & \text{if penny } j \text{ is to be left off the balance} \\ 1, & \text{if penny } j \text{ is to be placed in the right pan} \end{cases}$$

in the ith weighing. The four-penny problem may thus be restated as follows: For what *weighing design*, that is, for what 4×4 (nonsingular) design matrix $\boldsymbol{X} = (x_{ij})$ having for its elements only 1's, 0's, and -1's, is

$$\sum_{1 \leq i \leq 4} \text{Var}(\hat{\beta}_i)$$

the smallest? Solve the problem in this form.
[Hint: Use (2.8) and (1.8).] ◇

2.18 Exercise [2]. Consider the basis form (1.7) of the GLM. Let (c_{ij}) be defined by (1.13). Deduce from (1.12) that the GME of $\beta_i = \psi_i(\mu)$ is

$$\hat{\beta}_i = \sum_j c_{ij} \langle x_j, Y \rangle \quad \text{for } 1 \leq i \leq p, \tag{2.19}$$

from (2.7) that

$$E\hat{\beta}_i = \beta_i \quad \text{for } 1 \leq i \leq p, \tag{2.20}$$

and from (1.15) and (2.10) that

$$\text{Cov}(\hat{\beta}_i, \hat{\beta}_j) = \sigma^2 c_{ij} \quad \text{for } 1 \leq i, j \leq p. \tag{2.21}$$

The $\hat{\beta}_i$'s are called the *coefficients of the regression of Y on $x_1, \ldots, x_p$*. (Since $\hat{\beta}_i$ is the difference between the GMEs of $\sum_j (w_j + \delta_{ij}) \beta_j$ and $\sum_j w_j \beta_j$ for any constants $w_1, \ldots, w_p$, it is sometimes said that $\hat{\beta}_i$ estimates the effect of a unit increase in the ith predictor, the other predictors being held fixed.) ◇

Specialized to the standard formation

$$E(\boldsymbol{Y}) = \boldsymbol{X}\beta, \quad \Sigma_Y = \sigma^2 \boldsymbol{I}$$

of the GLM in $\mathbb{R}^n$, (2.19) says that

$$\hat{\beta} \equiv (\hat{\beta}_1, \ldots, \hat{\beta}_p)^T = \boldsymbol{C}\boldsymbol{X}^T\boldsymbol{Y} \tag{2.22}$$

with $C = (c_{ij})$, (2.20) says that

$$E(\hat{\beta}) = \beta, \tag{2.23}$$

and (2.21) says that

$$\Sigma_{\hat{\beta}} = \sigma^2 C. \tag{2.24}$$

2.25 Exercise [2]. As in Exercise 2.18, let $\hat{\beta}_1, \ldots, \hat{\beta}_p$ be the coefficients for the regression of Y on $x_1, \ldots, x_p$ (so $P_{[x_1,\ldots,x_p]}Y = \sum_{j=1}^{p} \hat{\beta}_j x_j$). For a given $n < p$, let $\hat{\beta}_1^*, \ldots, \hat{\beta}_n^*$ be the coefficients for the regression of Y on $x_1, \ldots, x_n$ (so $P_{[x_1,\ldots,x_n]}Y = \sum_{i=1}^{n} \hat{\beta}_i^* x_i$). For $i = 1, \ldots, n$, give a formula for $\hat{\beta}_i^*$ in terms of the $\hat{\beta}_j$'s and some additional regression coefficients, and discuss the statistical interpretation of that formula. [Hint: See Exercise 2.2.20.] $\qquad \diamond$

3. Estimation of μ itself

One can reformulate the preceding discussion to apply directly to the estimation of μ itself. Consider $P_M Y$. This is a linear unbiased estimator of μ since $E_\mu(P_M Y) = P_M \mu = \mu$ for all $\mu \in M$. Moreover, it has minimum dispersion in the class of linear unbiased estimators, in the following sense: For any linear unbiased estimator DY of μ, the dispersion operators of DY and $P_M Y$ satisfy

$$\Sigma(DY) \geq \Sigma(P_M Y), \tag{3.1}$$

the ordering relation

$$B \geq A$$

for self-adjoint linear transformations mapping V into V being interpreted as in (2.2.12) to mean

$$\langle x, Bx \rangle \geq \langle x, Ax \rangle \quad \text{for all } x \in V.$$

In view of the definition of dispersion operators, (3.1) reads simply

$$\text{Var}(\langle x, DY \rangle) \geq \text{Var}(\langle x, P_M Y \rangle) \quad \text{for all } x \in V; \tag{3.2}$$

this holds because $\langle x, DY \rangle$ is a linear unbiased estimator of its expected value $\langle x, D\mu \rangle = \langle x, \mu \rangle$ and so has a variance at least as large as that of the corresponding GME $\langle x, P_M Y \rangle$. Moreover, by the uniqueness part of the linear functional Gauss-Markov theorem, equality holds in (3.1) if and only if $\langle x, DY \rangle = \langle x, P_M Y \rangle$ for all x, that is, if and only if $DY = P_M Y$.

3.3 Exercise [3]. Suppose Ψ is a linear transformation mapping M into some inner product space, say $(W, \langle \cdot, \cdot \rangle_W)$. Show that $\Psi(P_M Y)$ is the *best linear unbiased estimator* of $\Psi(\mu)$, in the sense that it has minimum dispersion among unbiased estimators of $\Psi(\mu)$ of the form BY, where $B: V \to W$ is linear. Point out how this result encompasses both the "linear functional" and "vector" versions of the Gauss-Markov theorem. $\qquad \diamond$

3.4 Exercise [3]. Suppose $V = \mathbb{R}^n$ and that $e = (1, \ldots, 1)^T \in M$. Write $\boldsymbol{Y} = (Y_1, \ldots, Y_n)^T$.
(1) Show that the *fitted values* $\hat{Y}_i = (P_M \boldsymbol{Y})_i$, $1 \le i \le n$, have the same sample mean as the *observed values* Y_i, $1 \le i \le n$:

$$\frac{1}{n} \sum_{1 \le i \le n} \hat{Y}_i = \frac{1}{n} \sum_{1 \le i \le n} Y_i \equiv \bar{Y}. \tag{3.5}$$

(2) Prove the addition rule

$$V_T = V_E + V_R, \tag{3.6}$$

where

$$V_T = \sum_{1 \le i \le n} (Y_i - \bar{Y})^2 = \|Q_{[e]} \boldsymbol{Y}\|^2, \tag{3.7$_T$}$$

$$V_E = \sum_{1 \le i \le n} (\hat{Y}_i - \bar{Y})^2 = \|Q_{[e]} \hat{\boldsymbol{Y}}\|^2, \tag{3.7$_E$}$$

and

$$V_R = \sum_{1 \le i \le n} (Y_i - \hat{Y}_i)^2 = \|\boldsymbol{Y} - \hat{\boldsymbol{Y}}\|^2 \tag{3.7$_R$}$$

are, respectively, the so-called *total variation* (of the Y_i's about their mean), *explained variation* (that is, the variation in the Y_i's "explained" by the model assumption $E(Y) \in M$), and *residual variation* of $\boldsymbol{Y}$. (3) Show that

$$\frac{V_E}{V_T} = R^2, \tag{3.8}$$

where

$$R = \frac{\sum_i (Y_i - \bar{Y})(\hat{Y}_i - \bar{Y})}{\left(\sum_i (Y_i - \bar{Y})^2\right)^{1/2} \left(\sum_i (\hat{Y}_i - \bar{Y})^2\right)^{1/2}} = \frac{\langle Q_{[e]} \boldsymbol{Y}, Q_{[e]} \hat{\boldsymbol{Y}} \rangle}{\|Q_{[e]} \boldsymbol{Y}\| \, \|Q_{[e]} \hat{\boldsymbol{Y}}\|} \tag{3.9}$$

is the sample correlation coefficient between the observed and fitted values. (In practice, the larger is R^2, the better.) (4) Show that $R \ge 0$. (5) Finally, show that the fitted values $\hat{Y}_i$ are uncorrelated with the *residuals* $Y_j - \hat{Y}_j$:

$$\mathrm{Cov}(\hat{Y}_i, Y_j - \hat{Y}_j) = 0 \quad \text{for } 1 \le i \le n, \, 1 \le j \le n. \tag{3.10}$$

[Hint: Note that $Q_{[e]} P_M = P_{M-[e]}$.]   ⋄

To determine $P_M Y = \hat{\mu}$, one may call upon the various results about projections developed in Sections 1, 2, and 3 of Chapter 2. For example, one may express $P_M Y$:
(1) in terms of an orthogonal basis for M (see (2.1.14)), perhaps obtained by a Gram-Schmidt orthogonalization (see (2.1.15)); (2) in terms of the solution to an appropriate set of normal equations (see (2.1.26)); as the difference (Proposition 2.2.13) or product (Proposition 2.2.29) of known projections onto subspaces suitably related to M; or, especially, as the sum (Proposition 2.2.24) of projections onto mutually orthogonal subspaces of M, perhaps obtained with the aid of Tjur's theorem (Theorem 2.3.2).

3.11 Example. Suppose

$$M = \sum_{L \in \mathcal{T}} L, \tag{3.12}$$

where $\mathcal{T}$ is a subset of a Tjur system $\mathcal{L}$ of mutually book orthogonal subspaces of V. By Tjur's theorem,

$$M = \sum_{L \in \mathcal{T}} \left(\sum_{K \leq L} K^\circ \right) = \sum_{K \in \mathcal{T}^*} K^\circ, \tag{3.13}$$

where

$$\mathcal{T}^* = \{ K \in \mathcal{L} : K \leq L \text{ for some } L \in \mathcal{T} \} \tag{3.14}$$

and $K^\circ = K - \sum_{J \in \mathcal{L}, J < K} J$. Because the K°'s are mutually orthogonal,

$$\hat{\mu} = P_M Y = \sum_{K \in \mathcal{T}^*} P_{K^\circ} Y. \tag{3.15}$$

Section 2.3 discussed how the P_{K°'s can be calculated from the P_K's with the aid of a structure diagram for $\mathcal{L}$.

For example, if $\mathcal{L}$ is the Tjur system $\{V, R, G, C\}$ of Example 2.3.18 and $M = R + C$, so $\mathcal{T} = \{R, C\}$ and $\mathcal{T}^* = \{R, C, G\}$, then

$$\begin{aligned}
\hat{\mu} &= P_{R^\circ} Y + P_{C^\circ} Y + P_{G^\circ} Y \\
&= P_{R-G} Y + P_{C-G} Y + P_G Y = P_R Y + P_C Y - P_G Y,
\end{aligned} \tag{3.16}$$

in agreement with Exercise 2.2.26. •

3.17 Exercise [3]. Use (3.16) to derive (2.16) anew. Begin by checking that the regression manifold M in Example 3.11 is the same as that in Exercise 2.14. ◇

3.18 Example. In the framework of (1.7), one can determine

$$\hat{\mu} = \sum_{1 \leq j \leq p} \psi_j(\hat{\mu}) x_j \equiv \sum_{1 \leq j \leq p} \hat{\beta}_j x_j$$

by solving for $\hat{\beta}_1, \ldots, \hat{\beta}_p$ in the *normal equations*

$$\langle x_i, Y \rangle = \sum_{1 \leq j \leq p} \langle x_i, x_j \rangle \hat{\beta}_j, \quad 1 \leq i \leq p, \tag{3.19}$$

(see (2.1.26)) to get

$$\hat{\beta}_i = \sum_{1 \leq j \leq p} c_{ij} \langle x_j, Y \rangle, \tag{3.20}$$

in agreement with (2.19). •

Once $\hat{\mu}$ is in hand, the GME of any given linear functional ψ of μ is of course readily available. On the other hand, when one is able to find the GMEs of linear functionals directly, as in the previous section, the result of the following exercise may well be the easiest route to $\hat{\mu}$.

3.21 Exercise [1]. Suppose each x in M can be written in the form

$$x = \sum_{1 \leq i \leq k} \psi_i(x) x_i$$

for given linear functionals $\psi_i \colon M \to \mathbb{R}$ and (possibly linearly dependent) vectors $x_i \in M$, $1 \leq i \leq k$. Show that

$$\hat{\mu} = \sum_{1 \leq i \leq k} \hat{\psi}_i x_i, \tag{3.22}$$

$\hat{\psi}_i$ being the GME of $\psi_i(\mu)$. ◇

4. Estimation of σ^2

By Theorem 3.8.2, the residual vector $Q_M Y$ has in $M^\perp$ a weakly spherical distribution with zero mean and dispersion operator $\sigma^2 I_{M^\perp}$. Thus, when $d(M) < d(V)$,

$$\hat{\sigma}^2 \equiv \frac{\|Q_M Y\|^2}{d(M^\perp)} = \frac{\|Y\|^2 - \|P_M Y\|^2}{d(V) - d(M)} \tag{4.1}$$

unbiasedly estimates σ^2. The customary estimator of the standard deviation $\sigma \|\psi\|$ of the GME $\hat{\psi}(Y)$ of a linear functional $\psi(\mu)$ is

$$\hat{\sigma}_{\hat{\psi}} \equiv \hat{\sigma} \|\psi\| = \hat{\sigma} \|cv(\psi)\|. \tag{4.2}$$

4.3 Example. Consider simple linear regression: $V = \mathbb{R}^n$, $Y_1, \ldots, Y_n$ are uncorrelated random variables with equal variances σ^2 and $E(Y_i) = \alpha + \beta(x_i - \bar{x})$ for $1 \leq i \leq n$ with $x_1, \ldots, x_n$ known constants; note that the parameterization used here is different from that used in Examples 1.1.2 and 1.9. The regression manifold M is

$$[e, v],$$

where $(e)_i = 1$ and $(v)_i = x_i - \bar{x}$ for each i. Because $e \perp v$, one has

$$P_M Y = \hat{\alpha} e + \hat{\beta} v,$$

where

$$\hat{\alpha} = \left\langle \frac{e}{\|e\|^2}, Y \right\rangle = \bar{Y} \quad \text{and} \quad \hat{\beta} = \left\langle \frac{v}{\|v\|^2}, Y \right\rangle = \frac{\sum_i (x_i - \bar{x}) Y_i}{\sum_i (x_i - \bar{x})^2}. \tag{4.4}$$

Hence

$$\hat{\sigma}^2 = \frac{\|Q_M Y\|^2}{n - 2} \tag{4.5}$$

with

$$\|Q_M Y\|^2 = \|Y - (\hat{\alpha} e + \hat{\beta} v)\|^2 = \sum_i \left(Y_i - \hat{\alpha} - \hat{\beta}(x_i - \bar{x}) \right)^2 \tag{4.6}$$

$$= \|Y\|^2 - \|\hat{\alpha} e\|^2 - \|\hat{\beta} v\|^2 = \sum_i Y_i^2 - n \hat{\alpha}^2 - \hat{\beta}^2 \sum_i (x_i - \bar{x})^2. \tag{4.7}$$

The sum of squares on the far right of (4.6) is called a *closed form,* whereas the sum of squares on the far right of (4.7), which can be obtained algebraically by "opening" (that is, expanding) $(Y_i - \hat{\alpha} - \hat{\beta}(x_i - \bar{x}))^2$ and summing over i, is called an *open form.* The estimators of the standard deviations of $\hat{\alpha}$ and $\hat{\beta}$ are, respectively,

$$\hat{\sigma}_{\hat{\alpha}} = \hat{\sigma}/\|e\| = \hat{\sigma}/\sqrt{n} \tag{4.8}$$

and

$$\hat{\sigma}_{\hat{\beta}} = \hat{\sigma}/\|v\| = \hat{\sigma}/\sqrt{\sum_i (x_i - \bar{x})^2} \,. \tag{4.9} \bullet$$

4.10 Exercise [1]. What are the open and closed forms of $\|Q_M Y\|^2$ in the one-way analysis of variance layout of Example 2.2(D)?
[Hint: See Exercise 2.2.14.] $\diamond$

4.11 Example. Suppose that, as in Example 3.11,

$$M = \sum\nolimits_{L \in \mathcal{T}} L = \sum\nolimits_{K \in \mathcal{T}^*} K^{\circ},$$

where $\mathcal{T}$ is a subset of a Tjur system $\mathcal{L}$ and $\mathcal{T}^*$ consists of the elements of $\mathcal{L}$ that are at or below the level of some $L \in \mathcal{T}$ in the structure diagram for $\mathcal{L}$. Because the mutually orthogonal subspaces L° for $L \in \mathcal{L}$ decompose V, one has

$$\|Q_M Y\|^2 = \left\|\sum\nolimits_{K \notin \mathcal{T}^*} P_{K^{\circ}} Y\right\|^2 = \sum\nolimits_{K \notin \mathcal{T}^*} \|P_{K^{\circ}} Y\|^2; \tag{4.12}$$

the open-form expression on the far right of (4.12) is easily evaluated by reference to the structure diagram and analysis of variance table for $\mathcal{L}$. $\bullet$

4.13 Exercise [2]. Suppose the running assumption that $\mu \in M$ (see (0.1)) is false and that in fact μ lies in given subspace N of V that contains M. What then is the expected value of the estimator $\hat{\sigma}^2$ defined by (4.1)? $\diamond$

4.14 Exercise [3]. Put $q = d(M^{\perp})$ and suppose $\langle v_1, Y \rangle, \ldots, \langle v_q, Y \rangle$ are uncorrelated linear functionals of Y such that

$$E_{\mu}(\langle v_k, Y \rangle) = 0 \quad \text{and} \quad \text{Var}(\langle v_k, Y \rangle) > 0$$

for each $\mu \in M$ and each $1 \le k \le q$. Show that

$$\|Q_M Y\|^2 = \sum\nolimits_{1 \le k \le q} \frac{\langle v_k, Y \rangle^2}{\|v_k\|^2}. \tag{4.15} \diamond$$

4.16 Exercise [2]. Suppose $V = \mathbb{R}^n$ and M is spanned by the columns of a full rank matrix X. Show that the closed-form expression

$$(Y - X\hat{\beta})^T (Y - X\hat{\beta}) \tag{4.17}$$

for $\|Q_M Y\|^2$ has the equivalent open form

$$Y^T Y - \hat{\beta}^T X^T X \hat{\beta} = Y^T Y - \hat{\beta}^T (X^T Y). \tag{4.18} \diamond$$

4.19 Exercise [1]. Suppose $\mathcal{L}$ is the Tjur system $\{V, R, C, G\}$ of Example 2.3.18 and $M = R + C$. What are the closed- and open-form expressions for $\|Q_M Y\|^2$? [Hint: See Table 2.3.19.]   ◇

5. Using the wrong inner product

On occasion one may, out of ignorance or for the sake of simplicity, employ the wrong inner product in estimating μ, that is, one may estimate μ by $P_M^* Y$, where P_M^* denotes projection onto M with respect to an inner product

$$\langle \cdot, \cdot \rangle^* = \langle \cdot, \Delta \cdot \rangle = \langle \cdot, \cdot \rangle_\Delta \tag{5.1}$$

different from $\langle \cdot, \cdot \rangle$, $\Delta : V \to V$ being a positive-definite self-adjoint linear transformation. Consider, for example, the situation where $V = \mathbb{R}^n$, $\langle \cdot, \cdot \rangle^*$ is the dot-product, and $\langle \cdot, \cdot \rangle = \langle \cdot, B^{-1} \cdot \rangle^*$, $\sigma^2 B$ being the dispersion operator of Y with respect to $\langle \cdot, \cdot \rangle^*$; in this case, $P_M^* Y$ is the *least squares estimator* of μ, whereas $P_M Y$ is the GME of μ.

Now, because $E_\mu(P_M^* Y) = P_M^* \mu = \mu$ for all $\mu \in M$, P_M^* is a linear unbiased estimator of μ and so, by (3.1),

$$\Sigma(P_M^* Y) \geq \Sigma(P_M Y), \tag{5.2}$$

with equality if and only if

$$P_M^* Y = P_M Y. \tag{5.3}$$

Here we address the question of when (5.3) holds. The deeper problem of how much larger $\Sigma(P_M^* Y)$ is than $\Sigma(P_M Y)$ when $P_M^* Y \neq P_M Y$ is explored in the problem set at the end of this chapter.

We need to make a preliminary observation, namely, that for any linear transformation T mapping V into V, the adjoint T' of T with respect to $\langle \cdot, \cdot \rangle$ and the adjoint T'^* of T with respect to $\langle \cdot, \cdot \rangle^*$ are related by the *adjoint identities*

$$T'^* = \Delta^{-1} T' \Delta \quad \text{and} \quad \Delta T'^* \Delta^{-1} = T' \tag{5.4}$$

because, for example,

$$\langle x, Ty \rangle^* = \langle \Delta x, Ty \rangle = \langle T' \Delta x, y \rangle = \langle \Delta \Delta^{-1} T' \Delta x, y \rangle = \langle \Delta^{-1} T' \Delta x, y \rangle^*$$

for all $x, y \in V$.

5.5 Proposition. *Let Δ, P_M, and P_M^* be as above. The following are equivalent:*

(i) $P_M = P_M^*$,

(ii) P_M^* *and Δ commute:* $P_M^* \Delta = \Delta P_M^*$,

(iii) M *is invariant under Δ:* $\Delta(M) \subset M$,

(iv) M *admits a basis of eigenvectors of Δ,*

and imply that the "false" GME $P_M^* Y$ of μ coincides with the true GME $P_M Y$.

The proposition remains valid with P_M^* replaced by P_M in (ii). However, we have stated (ii) in terms of P_M^* because it is presumably P_M^* that one has in hand, since one is doing the computations with respect to $\langle\cdot,\cdot\rangle^*$. In interpreting the proposition, bear in mind that since $\sigma^2 I$ is by assumption the dispersion operator of Y in the $\langle\cdot,\cdot\rangle$ inner product, $\sigma^2\Delta = \sigma^2 I\Delta$ is the dispersion operator of Y in the $\langle\cdot,\cdot\rangle^*$ inner product.

Proof of Proposition 5.5. (i) *implies* (ii): P_M^* is $\langle\cdot,\cdot\rangle^*$-self-adjoint, so by the adjoint identity (5.4),

$$(P_M^*)' = \Delta\big((P_M^*)'^*\big)\Delta^{-1} = \Delta P_M^*\Delta^{-1}.$$

Given that $P_M^* = P_M$, P_M^* is also $\langle\cdot,\cdot\rangle$-self-adjoint, whence $(P_M^*)' = P_M^*$ and $P_M^*\Delta = \Delta P_M^*$.

(ii) *implies* (iii): For each $x \in M$, one has $\Delta x = \Delta P_M^* x = P_M^* \Delta x \in M$.

(iii) *implies* (i): Supposing $\Delta(M) \subset M$, the identity

$$\langle v, m\rangle^* = \langle v, \Delta m\rangle \quad \text{for } v \in V,\, m \in M$$

tells us that $v \perp M$ implies $v \perp^* M$. For $x \in V$, the decomposition

$$x = P_M x + (x - P_M x)$$

is thus the (necessarily unique) representation of x as the sum of a vector (namely $P_M x$) in M and a vector (namely $x - P_M x$) $\langle\cdot,\cdot\rangle^*$-orthogonal to M, whence $P_M^* x = P_M x$.

(iii) *implies* (iv): Supposing again that $\Delta(M) \subset M$, we may view Δ as a linear transformation of M into itself. This restricted Δ is self-adjoint with respect to $\langle\cdot,\cdot\rangle$, restricted to $M \times M$. The Spectral theorem gives the existence of a $\langle\cdot,\cdot\rangle$-orthonormal basis $e_1,\ldots,e_p$ for M such that for appropriate scalars λ_i, $\Delta e_i = \lambda_i e_i$ for each i.

(iv) *implies* (iii): Trivial. ∎

5.6 Exercise [1]. (1) Show that an additional equivalent condition is

   (v) $M^\perp$ *is invariant under* Δ: $\Delta(M^\perp) \subset M^\perp$.

(2) Show that the proposition remains valid if in its statement Δ is replaced by Δ^{-1} and/or P_M^* is switched with P_M.
[Hint: No calculations are needed!]   ◇

5.7 Example. Take $V = \mathbb{R}^n$, $\langle\cdot,\cdot\rangle^* =$ dot-product, and $\langle\cdot,\cdot\rangle = \langle\cdot, B^{-1}\cdot\rangle^*$, where $\sigma^2 B$ is the dispersion operator of Y with respect to $\langle\cdot,\cdot\rangle^*$. Suppose $M = [e]$, so that one is dealing with the one-sample problem. The proposition says that the least squares estimator $P_M^* Y$ is the same as the GME $P_M Y$ if and only if e is an eigenvector of B. This is the case, for example, in the *equicorrelated* situation

$$\text{Correlation}(Y_i, Y_j) = \begin{cases} 1, & \text{if } i = j, \\ \rho, & \text{if } i \neq j; \end{cases} \tag{5.8}$$

here e is an eigenvector with eigenvalue $1 + (n-1)\rho$.   ●

5.9 Exercise [2]. Suppose that in Example 5.7 the matrix representing B is diagonal. Under what further conditions on B will $P_M Y = P_M^* Y$ when: (1) $M = [e]$; (2) M is given by (2.3)?   ◇

5.10 Exercise [2]. Suppose $P_M Y = P_M^* Y$. Show that the corresponding estimators

$$\frac{\|Y - P_M Y\|^2}{d(V) - d(M)} \quad \text{and} \quad \frac{(\|Y - P_M^* Y\|^*)^2}{d(V) - d(M)}$$

of σ^2 are the same if and only if the restriction of Δ to $M^\perp$ is the identity: $\Delta m^\perp = m^\perp$ for all $m^\perp \in M^\perp$. In the context of Example 5.7, for what value(s) of ρ in (5.8) does this latter condition hold?   ◇

5.11 Exercise [3]. It may be the case that $P_M Y$ and $P_M^* Y$ are not identical, but yet that the false GME $\psi(P_M^* Y)$ and the true GME $\psi(P_M Y)$ of a given linear functional ψ of μ are nonetheless the same. Let $c \in M$ be the coefficient vector of ψ with respect to $\langle \cdot, \cdot \rangle$ and let c^* be the coefficient vector of ψ with respect to $\langle \cdot, \cdot \rangle^*$. Show that the following are equivalent:

   (i) $\Delta c^* \in M$,

   (ii) $c = \Delta c^*$,

   (iii) $c^* = \Delta^{-1} c$,

   (iv) $\Delta^{-1} c \in M$,

and

   (v) $\psi P_M = \psi P_M^*$.

[Hint: assertion (v) says $\langle c, P_M x \rangle = \langle c, P_M^* x \rangle$ for all $x \in V$. By the adjoint identity (5.4), this is equivalent to $\langle c, x \rangle = \langle \Delta P_M^* \Delta^{-1} c, x \rangle$ for all $x \in V$ and hence to (iv).]   ◇

6. Invariance of GMEs under linear transformations

The following discussion shows that if one is given a nonsingular linear transformation T mapping V into V, it is permissible to carry out the GM estimation process by first transforming the problem via T, then doing GM estimation in the transformed problem, and finally transforming back via T^{-1}. Such an approach may be advantageous when projection is relatively easy to carry out in the transformed problem.

    Let then $T \colon V \to V$ be a nonsingular linear transformation. By (3.4.5), TY has dispersion operator TT' with respect to $\langle \cdot, \cdot \rangle$ and therefore is weakly spherical with respect to the inner product

$$\langle \cdot, \cdot \rangle^* = \langle \cdot, (TT')^{-1} \cdot \rangle. \tag{6.1}$$

Moreover,

$$E(TY) = T\mu \in TM \equiv \{Tm : m \in M\}.$$

Let projection onto TM with respect to $\langle \cdot, \cdot \rangle^*$ be denoted P_{TM}^*. By Exercise 6.3 below,

$$P_M = T^{-1} P_{TM}^* T. \tag{6.2}$$

From a vector point of view, (6.2) says that the GME $\hat{\mu} = P_M Y$ of μ may be obtained by applying T^{-1} to the GME $P^*_{TM} TY$ of $T\mu$ in the transformed problem. From a linear functional point of view, (6.2) says that the GME of a linear functional ψ on M is $\psi(P_M Y) = \psi T^{-1}(P^*_{TM} TY)$, the GME in the transformed problem of the linear functional ψT^{-1} on TM.

If ψ is the linear functional recording the coordinate of a vector in M with respect to the jth element x_j of a basis $x_1, \ldots, x_p$ for M, then ψT^{-1} records the coordinate of x in TM with respect to Tx_j, the jth element of the transformed basis $Tx_1, \ldots, Tx_p$, since

$$\psi T^{-1}\Big(\sum_i \beta_i Tx_i\Big) = \psi\Big(\sum_i \beta_i x_i\Big) = \beta_j.$$

According to the preceding discussion, the GME $\hat{\beta}_j$ of β_j can be found either by projecting Y onto M and reading off the jth coordinate with respect to the x_i's, or by projecting TY onto TM (using $\langle \cdot, \cdot \rangle^*$) and reading off the jth coordinate with respect to the Tx_i's.

6.3 Exercise [3]. Prove (6.2).
[Hint: That $T^{-1} P^*_{TM} T$ is $\langle \cdot, \cdot \rangle$-self-adjoint follows from the adjoint identity (5.4).] $\qquad$ ◇

To avoid confusion as to the role of the symbol T, for the rest of this section let the transpose of a vector $\boldsymbol{x} \in \mathbb{R}^n$ and a matrix $\boldsymbol{A}$ be denoted by $\boldsymbol{x}'$ and $\boldsymbol{A}'$, rather than the usual $\boldsymbol{x}^T$ and $\boldsymbol{A}^T$.

6.4 Example. Suppose $\boldsymbol{Y}$ is a random vector in $\mathbb{R}^n$ with covariance matrix of the form $\sigma^2 \boldsymbol{B}$ for a known positive-definite matrix $\boldsymbol{B}$. Then $\boldsymbol{Y}$ is weakly spherical with respect to the inner product $\langle \boldsymbol{x}, \boldsymbol{y} \rangle = \boldsymbol{x}' \boldsymbol{B}^{-1} \boldsymbol{y}$. Let $\boldsymbol{T}$ be the matrix representing the transformation T in the preceding discussion. By (6.1), $\langle \boldsymbol{x}, \boldsymbol{y} \rangle^* = \langle \boldsymbol{T}^{-1}\boldsymbol{x}, \boldsymbol{T}^{-1}\boldsymbol{y} \rangle = \boldsymbol{x}'(\boldsymbol{T}^{-1})'\boldsymbol{B}^{-1}\boldsymbol{T}^{-1}\boldsymbol{y} = \boldsymbol{x}'(\boldsymbol{T}\boldsymbol{B}\boldsymbol{T}')^{-1}\boldsymbol{y}$, so if $\boldsymbol{T}$ is chosen so that $\boldsymbol{T}\boldsymbol{B}\boldsymbol{T}' = \boldsymbol{I}$, then $\langle \cdot, \cdot \rangle^*$ is the well-understood and easily handled dot-product. $\qquad$ •

6.5 Exercise [2]. Continuing Example 6.4, show that such a $\boldsymbol{T}$ exists. $\qquad$ ◇

6.6 Example. Let $V = \mathbb{R}^n$, let $p = \dim(M)$, and let T be any orthogonal linear transformation of $\mathbb{R}^n$ into itself that carries the given manifold M into the "first p-dimensional" subspace $\{(x_1, \ldots, x_n)' \in \mathbb{R}^n : x_{p+1} = \cdots = x_n = 0\}$ of $\mathbb{R}^n$. Since T is orthogonal, the $\langle T^{-1} \cdot, T^{-1} \cdot \rangle$ inner product is the original dot-product, and $P^*_{TM}((x_1, \ldots, x_n)') = (x_1, \ldots, x_p, 0, \ldots, 0)'$ for all $\boldsymbol{x} \in \mathbb{R}^n$. The transformation T is traditionally called a *canonical transformation* and is said to *reduce the problem to canonical form;* it is often used as a theoretical tool when the GLM is treated from a coordinatized point of view. $\qquad$ •

7. Some additional optimality properties of GMEs

We consider first an optimality criterion of the GME $P_M Y$ outside the class of linear unbiased estimators of μ. We need a criterion to measure the performance of a not necessarily unbiased estimator $\hat{\mu}$ of μ. In the one-dimensional setting it is customary to employ the *mean square error (MSE)*

$$R(\hat{\mu}, \mu) = E_\mu (\hat{\mu} - \mu)^2 = \text{Var}_\mu(\hat{\mu}) + (E_\mu \hat{\mu} - \mu)^2. \tag{7.1}$$

To say $\hat{\mu}$ has a small MSE is to say that $\hat{\mu}$ has both a small standard deviation and a small bias.

In the vector setting, a natural definition of MSE is

$$R(\hat{\mu}, \mu) = E_\mu (\|\hat{\mu} - \mu\|^2). \tag{7.2}$$

Since $\hat{\mu} - \mu$ has dispersion operator I with respect to $\langle \cdot, \cdot \rangle^* \equiv \langle \cdot, \Sigma^{-1} \cdot \rangle$, Σ being the dispersion operator of $\hat{\mu}$, and since $\|\hat{\mu} - \mu\|^2 = \langle \Sigma(\hat{\mu} - \mu), \hat{\mu} - \mu \rangle^*$, formula (3.9.1) gives

$$R(\hat{\mu}, \mu) = \text{trace}(\Sigma) + \|E_\mu \hat{\mu} - \mu\|^2, \tag{7.3}$$

this being an analog to (7.1).

The Gauss-Markov theorem implies that $P_M Y$ has minimum MSE in the class of linear unbiased estimators of μ. Indeed, letting $e_1, \ldots, e_n$ be an orthonormal basis for V and assuming DY to be a linear unbiased estimator of μ, we have

$$R(DY, \mu) = E_\mu (\|DY - \mu\|^2) = \sum_i E_\mu (\langle DY - \mu, e_i \rangle^2)$$

$$= \sum_i \text{Var}_\mu (\langle DY, e_i \rangle) \geq \sum_i \text{Var}_\mu (\langle P_M Y, e_i \rangle)$$

$$= \sum_i E_\mu (\langle P_M Y - \mu, e_i \rangle^2) = R(P_M Y, \mu),$$

the inequality holding because $\langle DY, e_i \rangle$ is a linear unbiased estimator of its expected value $\langle D\mu, e_i \rangle = \langle \mu, e_i \rangle$, the GM estimator of which is $\langle P_M Y, e_i \rangle$.

7.4 Exercise [3]. Show that for any linear transformation $A\colon V \to V$, the quantity

$$\sum_{1 \leq i \leq n} \langle e_i, A e_i \rangle \tag{7.5}$$

does not depend on which orthonormal basis $e_1, \ldots, e_n$ is employed in V; the common value of all these sums is called trace(A). When A is self-adjoint, trace(A) is thus the sum of the eigenvalues of A, multiplicities included, as defined in part **A** of Section 3.9. Check that for self-adjoint transformations A and B,

$$A \leq B \quad \text{implies} \quad \text{trace}(A) \leq \text{trace}(B) \tag{7.6}$$

and use this together with (3.1) and (7.3) to give an alternate proof of the fact that $P_M Y$ has minimum MSE among linear unbiased estimators of μ. ◇

One can push these ideas a bit further to obtain

7.7 Proposition. $\hat{\mu} = P_M Y$ *has minimum MSE in the class of affine estimators of μ with bounded MSE.*

Proof. Let $DY + \mu_0$ be an *affine estimator* ($D: V \to V$ is linear, $\mu_0 \in V$) of μ with bounded MSE. Note that

$$
\begin{aligned}
R(DY + \mu_0, \mu) &= E_\mu \|DY + \mu_0 - \mu\|^2 \\
&= E_\mu \|(DY - D\mu) + (D\mu - \mu + \mu_0)\|^2 \\
&= E_\mu \|DY - D\mu\|^2 + 2E_\mu \langle DY - D\mu, D\mu - \mu + \mu_0 \rangle \\
&\quad + \|D\mu - \mu + \mu_0\|^2 \\
&= E_\mu \|D(Y - \mu)\|^2 + 0 + \|(D\mu - \mu) + \mu_0\|^2.
\end{aligned}
$$

Were $D\mu_1 \neq \mu_1$ for some $\mu_1 \in M$, $DY + \mu_0$ would have unbounded MSE (consider $\|(Dc\mu_1 - c\mu_1) + \mu_0\|^2$ as a function of $c \in \mathbb{R}$), so we must have

$$
D\mu = \mu \quad \text{for all } \mu \in M.
$$

But then DY is a linear unbiased estimator of μ, and the effect of translating it by μ_0 is only to make the bias, and so also the MSE, bigger:

$$
R(DY + \mu_0, \mu) = E_\mu \|DY - D\mu\|^2 + \|\mu_0\|^2 \geq R(DY, \mu) \geq R(P_M Y, \mu). \qquad \blacksquare
$$

Finally we consider an invariance argument that also leads to $P_M Y$ as the "ideal" estimator of μ. Consider a generic estimator $g(Y)$ of the mean $E(Y) \in M$ of our weakly spherical random vector Y. It is reasonable to ask that

$$
g(y) \in M \quad \text{for all } y \in V. \tag{7.8$_1$}
$$

Now for $m \in M$, the random vector $Y + m$ is weakly spherical with mean

$$
E(Y + m) = E(Y) + m \in M,
$$

and so it would be natural to use $g(Y + m)$ to estimate the left-hand side of the above equation. Since $g(Y) + m$ estimates the right-hand side, it is reasonable to ask also that

$$
g(y + m) = g(y) + m \quad \text{for all } y \in V \text{ and all } m \in M. \tag{7.8$_2$}
$$

Similarly, for each orthogonal transformation $\mathcal{O}: V \to V$ that maps M into M, the random vector $\mathcal{O}Y$ is weakly spherical with mean

$$
E(\mathcal{O}Y) = \mathcal{O}(E(Y)) \in M,
$$

and it is reasonable to ask also that

$$
g(\mathcal{O}y) = \mathcal{O}g(y) \quad \text{for all } y \in V \text{ and all such mappings } \mathcal{O}. \tag{7.8$_3$}
$$

7.9 Proposition. *The unique $g: V \to V$ satisfying* (7.8) *is* $g = P_M$.

7.10 Exercise [3]. Prove the proposition.
[Hint: First show that $g(y) = 0$ for each $y \in M^\perp$.]   ◇

8. Estimable parametric functionals

When $V = \mathbb{R}^n$ the regression manifold M is often written as

$$M = \{ \, \boldsymbol{X\beta} : \boldsymbol{\beta} \in \mathbb{R}^k \, \}, \tag{8.1}$$

where the $n \times k$ matrix $\boldsymbol{X}$ may or may not be of full rank. This situation can be abstracted to the case of a general V by stipulating that

$$M = \{ \, X\beta : \beta \in B \, \} \tag{8.2}$$

is the range of a linear transformation X mapping an inner product space B into V; X may or may not be one-to-one. Written out in full, the distributional assumptions on Y corresponding to (8.2) read

$$\begin{array}{c} \text{the random vector } Y \text{ is weakly spherical} \\ \text{in } V \text{ and has mean } \mu \,=\, E(Y) \text{ of the} \\ \text{form } X\beta \text{ for some unknown } \beta \text{ lying in } B. \end{array} \tag{8.3}$$

The parameter β in (8.3) is of special interest. When X is one-to-one, $X'X$ is invertible so that $\beta = (X'X)^{-1}X'\mu$ is a linear function of $\mu = X\beta$. By the Gauss-Markov theorems, β and all linear functionals of it accordingly have best linear unbiased estimators. On the other hand, when X is not one-to-one, various structural difficulties arise. β itself is not uniquely determined by μ and so has no natural estimator whatsoever. The same is true of some linear functionals of β. As it turns out, other linear functionals of β are uniquely determined by μ, and in fact linearly so, and therefore do have best linear unbiased estimators. This section delineates which functionals of β are so estimable and what their optimal estimators are. Until further notice we adopt (8.3) as the running assumption and refer to (8.1) and (8.2) as the "matrix case" and "general case," respectively. We will call an arbitrary linear functional of β a *parametric functional*. In the matrix case a parametric functional can be thought of as a linear combination $p_1\beta_1 + \cdots + p_k\beta_k$ of the coordinates $\beta_1, \ldots, \beta_k$ of β.

8.4 Proposition. *Let ϕ be a parametric functional with coefficient vector p, so that $\phi(\beta) = \langle p, \beta \rangle_B$ for all $\beta \in B$. The following are equivalent:*

  (i) *$\phi(\beta)$ admits a linear unbiased estimator in the sense that there exists a $v \in V$ such that $E_\beta \langle v, Y \rangle_V = \phi(\beta)$ for all possible values of the parameter β in (8.3),*

 (ii) *there exists a linear functional ψ on M such that*

$$\phi(\beta) = \psi(X\beta) \quad \text{for all } \beta \in B, \tag{8.5}$$

 (iii) *$\phi(\beta)$ is uniquely determined by $X\beta$ in the sense that $\phi(\beta_1) = \phi(\beta_2)$ whenever $\beta_1, \beta_2 \in B$ satisfy $X\beta_1 = X\beta_2$,*

 (iv) *$p \perp \mathcal{N}(X)$,*

  (v) *$p \in \mathcal{R}(X')$,*

 (vi) *$p \in \mathcal{R}(X'X)$.*

Proof. (i) *implies* (ii): If $\phi(\beta) = E_\beta \langle v, Y \rangle_V$, then $\phi(\beta) = \langle v, E_\beta Y \rangle_V = \langle v, X\beta \rangle_V = \psi(X\beta)$ for $\psi = \langle v, \cdot \rangle_V$.

(ii) *implies* (iii): Trivial.

(iii) *implies* (iv): Suppose $\beta \in \mathcal{N}(X)$, so that $X\beta = 0 = X0$. Condition (iii) (with $\beta_1 = \beta$ and $\beta_2 = 0$) then implies that $\phi(\beta) = \phi(0) = 0$, so that $p \perp \beta$.

(iv) $\Longleftrightarrow$ (v) $\Longleftrightarrow$ (vi): By Exercises 2.5.9, 2.5.17, and 2.5.20, $\mathcal{N}^\perp(X) = \mathcal{R}(X') = \mathcal{R}(X'X)$.

(v) *implies* (i): Suppose $p = X'v$ for some $v \in V$. Then $E_\beta \langle v, Y \rangle_V = \langle v, E_\beta Y \rangle_V = \langle v, X\beta \rangle_V = \langle X'v, \beta \rangle_B = \langle p, \beta \rangle_B = \phi(\beta)$. ∎

A parametric functional $\phi(\beta)$ is said to be *(linearly and unbiasedly) estimable* if ϕ satisfies any of the equivalent conditions of Proposition 8.4; otherwise, $\phi(\beta)$ is said to be *nonestimable*. By (8.5), the estimable parametric functionals are precisely the ones that are linear functionals of the mean of Y.

8.6 Exercise [2]. Suppose $\phi(\beta)$ is an estimable parametric functional with coefficient vector p. Show that the coefficient vector of the linear functional ψ in (8.5) is the unique element m in M such that $p = X'm$.   ◇

8.7 Exercise [3]. Show that if the parametric functional $\phi(\beta)$ is nonestimable, then it is completely undetermined by $\mu = X\beta$, in the sense that for each $\mu \in M$ and each $c \in \mathbb{R}$, there exists a $\beta \in B$ such that $X\beta = \mu$ and $\phi(\beta) = c$.   ◇

It is worth spelling out what conditions (iv)–(vi) of Proposition 8.4 have to say about the estimability of $\boldsymbol{p}^T\boldsymbol{\beta}$ in the matrix case (8.1). Condition (iv) requires $\boldsymbol{p}^T\boldsymbol{b} = 0$ for all $\boldsymbol{b} \in \mathbb{R}^k$ such that $\boldsymbol{X}\boldsymbol{b} = 0$. Condition (v) requires $\boldsymbol{p}$ to be a linear combination of the columns of $\boldsymbol{X}^T$, that is, that $\boldsymbol{p}^T$ be a linear combination of the rows of $\boldsymbol{X}$. Since multiplication of the ith row of $\boldsymbol{X}$ into $\boldsymbol{\beta}$ gives the expected value of Y_i, this result says that the only linear combinations of $\beta_1, \ldots, \beta_k$ that are estimable are the ones that are linear combinations of the coordinates of $\boldsymbol{\mu} = \boldsymbol{X}\boldsymbol{\beta}$. Finally, condition (vi) requires that $\boldsymbol{p}$ be a linear combination of the columns of $\boldsymbol{X}^T\boldsymbol{X}$ or, equivalently, that $\boldsymbol{p}^T$ be a linear combination of the rows of that symmetric matrix.

8.8 Example *(The one-way layout).* Consider the one-way layout parameterized in the form

$$E(Y_{ij}) = \beta_0 + \beta_i \quad \text{for } 1 \le j \le n_i,\, 1 \le i \le I$$

with no restrictions on the β_i's. One thinks of β_0 as what the expected value of each Y_{ij} would be in the absence of any treatments and of β_i as an incremental effect due to the treatment applied to the ith group. A linear combination $p_0\beta_0 + \cdots + p_I\beta_I$ of $\beta_0, \ldots, \beta_I$ is estimable if and only if it can be written as a linear combination $c_1(\beta_0 + \beta_1) + \cdots + c_I(\beta_0 + \beta_I)$ of the coordinates $E(Y_{ij})$ of $\boldsymbol{\mu}$. A moment's reflection shows that the condition for estimability is that

$$p_0 = \sum_{1 \le i \le I} p_i. \tag{8.9}$$

Thus no single β_i is estimable — without further assumptions on the β_i's, it is impossible to disentangle the background level β_0 from the incremental effects $\beta_1, \ldots, \beta_I$. Moreover, a linear combination $\sum_{1 \le i \le I} p_i \beta_i$ of the incremental effects is estimable if and only if $\sum_{1 \le i \le I} p_i = 0$; for example, $\beta_2 - \beta_1$ and $\beta_3 - (\beta_1 + \beta_2)/2$ are estimable, but $(\beta_1 + \beta_2 + \beta_3)/3$ is not. •

8.10 Exercise [3]. Consider the additive two-way layout

$$E(Y_{ij}) = \nu + \alpha_i + \beta_j, \quad 1 \le i \le I, 1 \le j \le J,$$

with no restrictions on the parameters $\nu, \alpha_1, \ldots, \alpha_I, \beta_1, \ldots, \beta_J$. Show that a linear combination

$$\sum_{1 \le i \le I} p_i \alpha_i + \sum_{1 \le j \le J} q_j \beta_j + r\nu$$

is estimable if and only if

$$\sum_{1 \le i \le I} p_i = r = \sum_{1 \le j \le J} q_j. \qquad \diamond$$

8.11 Exercise [2]. Suppose $V = \mathbb{R}^n$. In the model

$$E(Y_i) = \beta_i - \beta_{i-1}, \quad 1 \le i \le n,$$

what linear combinations of the parameters $\beta_0, \ldots, \beta_n$ are estimable? $\diamond$

Returning to the general case (8.2), suppose $\phi(\beta)$ is an estimable parametric functional. As in (8.5), let ψ be a linear functional on M such that $\phi(\beta) = \psi(X\beta)$ for all $\beta \in B$; ψ is uniquely determined by ϕ because X maps B onto M. The *Gauss-Markov estimator* $\hat{\phi}(Y)$ *of* $\phi(\beta)$ is defined to be the GME $\hat{\psi}(Y) = \psi(P_M Y)$ of $\psi(EY)$.

8.12 Proposition (Gauss-Markov theorem for estimable functionals). *The best linear unbiased estimator of an estimable parametric functional $\phi(\beta)$ is its GME $\hat{\phi}(Y)$.*

Proof. In the framework of (8.3), a linear functional LY of Y is unbiased for $\phi(\beta)$ if and only if, for all $\beta \in B$ and all possible choices of the distribution of Y as being weakly spherical with mean $X\beta$, one has

$$E(LY) = \phi(\beta). \tag{8.13}$$

Because $\phi(\beta) = \psi(X\beta)$ for all β, the requirements are that for all $\mu \in M$ and all choices of the distribution of Y as being weakly spherical with mean μ, one has

$$E(LY) = \psi(\mu). \tag{8.14}$$

By the Gauss-Markov theorem for ψ, the choice of L minimizing $\text{Var}(LY)$ subject to the constraint (8.14), and hence subject to the equivalent constraint (8.13), is $L = \psi P_M$. ∎

8.15 Exercise [2]. Show that a nonestimable parametric functional $\phi(\beta)$ does not admit any unbiased estimator, linear or not.
[Hint: Specifically, show that there is no function $f: V \to \mathbb{R}$ such that $Ef(Y) = \phi(\beta)$ for all $\beta \in B$ and all possible choices of the distribution of Y as being weakly spherical with mean $X\beta$.]   ◇

8.16 Proposition. Let $\phi(\beta) = \langle p, \beta \rangle_B$ be an estimable parametric functional. Its GME $\hat{\phi}(Y)$ can be written variously as:

   (i) $\langle P_M v, Y \rangle_V$, for any $v \in V$ such that $E_\beta \langle v, Y \rangle_V = \phi(\beta)$ for all $\beta \in B$,

   (ii) $\langle P_M v, Y \rangle_V$, for any $v \in V$ such that $p = X'v$,

   (iii) $\phi(\hat{\beta})$, for any solution $\hat{\beta}$ to the normal equations $X'X\hat{\beta} = X'Y$, or

   (iv) $\langle b, X'Y \rangle_B$, for any $b \in B$ such that $p = X'Xb$.

Proof. (i): By condition (i) of Proposition 8.4 there is at least one $v \in V$ such that $\langle v, Y \rangle_V$ unbiasedly estimates $\phi(\beta)$. The first step of the proof of that proposition showed that for any such v, one has $\phi(\beta) = \psi(X\beta)$ for $\psi = \langle v, \cdot \rangle_V$. Hence $cv(\psi) = P_M v$ and $\hat{\phi}(Y) \equiv \hat{\psi}(Y) = \langle P_M v, Y \rangle_V$.

(ii): By condition (v) of Proposition 8.4 there is at least one $v \in V$ such that $p = X'v$. The last step of the proof of that proposition showed that for any such v, $\langle v, Y \rangle_V$ unbiasedly estimates $\phi(\beta)$, so (ii) follows from (i).

(iii): By Exercise 2.5.22 there is at least one solution $\hat{\beta}$ to the normal equations, and any such $\hat{\beta}$ satisfies $X\hat{\beta} = P_M Y$. Let v be as in (ii), so that $p = X'v$. Then $\hat{\phi}(Y) = \langle v, P_M Y \rangle_V = \langle v, X\hat{\beta} \rangle_V = \langle X'v, \hat{\beta} \rangle_B = \langle p, \hat{\beta} \rangle_B = \phi(\hat{\beta})$.

(iv): By condition (vi) of Proposition 8.4 there is at least one $b \in B$ such that $p = X'Xb$. For any such b and any solution $\hat{\beta}$ to the normal equations, one has $\langle b, X'Y \rangle_B = \langle b, X'X\hat{\beta} \rangle_B = \langle X'Xb, \hat{\beta} \rangle_B = \langle p, \hat{\beta} \rangle_B = \phi(\hat{\beta})$, so (iv) follows from (iii).∎

   In the matrix case, assertion (iii) of Proposition 8.16 says that the GME of $\boldsymbol{p}^T\boldsymbol{\beta}$ is $\boldsymbol{p}^T\hat{\boldsymbol{\beta}}$, where $\hat{\boldsymbol{\beta}}$ is any solution to the normal equations $\boldsymbol{X}^T\boldsymbol{X}\hat{\boldsymbol{\beta}} = \boldsymbol{X}^T\boldsymbol{Y}$. If $\boldsymbol{p}^T$ can be recognized as a linear combination $\boldsymbol{b}^T\boldsymbol{X}^T\boldsymbol{X}$ of the rows of the coefficient matrix of the normal equations, then the GME of $\boldsymbol{p}^T\boldsymbol{\beta}$ can be easily obtained without solving the normal equations; indeed, assertion (iv) of the proposition gives the GME of $\boldsymbol{p}^T\boldsymbol{\beta}$ as $\boldsymbol{b}^T(\boldsymbol{X}^T\boldsymbol{Y})$. In general, though, it is no easier to find a $\boldsymbol{b}$ such that $\boldsymbol{X}^T\boldsymbol{X}\boldsymbol{b} = \boldsymbol{p}$ than it is to solve the normal equations for $\hat{\boldsymbol{\beta}}$.

8.17 Example *(The one-way layout).* In the context of Example 8.8, the GME of an estimable parametric functional $\phi(\boldsymbol{\beta}) = \sum_{0 \leq i \leq I} p_i \beta_i$ is

$$\hat{\phi}(Y) = \sum_{1 \leq i \leq I} p_i \bar{Y}_i = \langle v, Y \rangle_V,$$

where

$$v = \sum_{1 \leq i \leq I} \frac{p_i v_i}{n_i}$$

with $(v_i)_{i'j} = \delta_{ii'}$, since $v \in M$ and

$$E_{\boldsymbol{\beta}}\langle v, Y \rangle_V = \sum_{1 \le i \le I} p_i(\beta_0 + \beta_i) = \sum_{0 \le i \le I} p_i\beta_i = \phi(\boldsymbol{\beta})$$

by (8.9).   •

8.18 Exercise [3]. Find the GME of an arbitrary estimable $\phi(\beta)$ in (1) Exercise 8.10 and (2) Exercise 8.11.   ◇

8.19 Exercise [2]. Show that if $\phi_1(\beta) = \langle p_1, \beta \rangle_B$ and $\phi_2(\beta) = \langle p_2, \beta \rangle_B$ are estimable parametric functionals, with $p_1 = X'Xb_1$ and $p_2 = X'Xb_2$ for b_1 and $b_2 \in B$, then

$$\text{Cov}(\hat{\phi}_1(Y), \hat{\phi}_2(Y)) = \sigma^2 \langle Xb_1, Xb_2 \rangle_V = \sigma^2 \langle b_1, p_2 \rangle_B = \sigma^2 \langle p_1, b_2 \rangle_B.$$

[Hint: See Exercise 8.6.]   ◇

8.20 Exercise [2]. Show that if the vector space of estimable functionals for the model (8.2) is spanned by $\phi_1, \ldots, \phi_I$, then the regression manifold M is spanned by the coefficient vectors $m_1, \ldots, m_I$ of the corresponding GMEs $\hat{\phi}_i(Y) = \langle m_i, P_M Y \rangle_M$.   ◇

8.21 Example. As in Section 2.3, let $\mathcal{D}$ be a Tjur design consisting of orthogonal factors F (see (2.3.65)). Let V be the inner product space of real-valued functions on the set of experimental units, let $\mathcal{F}$ be a subset of $\mathcal{D}$, and let Y be a V-valued weakly spherical random vector with mean of the form

$$E(Y) = \sum_{F \in \mathcal{F}} \left(\sum_{f \in F} \beta_f^F I_f \right) \tag{8.22}$$

for some unknown constants β_f^F ($f \in F$, $F \in \mathcal{F}$); here I_f is the indicator function of f (see (2.3.47)). Equation (8.22) postulates that the factors in $\mathcal{F}$ are additive in their effects and that β_f^F is the effect of the fth level of the factor F.

This scenario is a particular case of (8.3): Take

$$B = \{ (\beta_f^F)_{f \in F, F \in \mathcal{F}} : \beta_f^F \in \mathbb{R} \text{ for each } f \in F \in \mathcal{F} \}$$

and let X be the linear transformation from B to V mapping $\beta \in B$ to

$$X\beta = \sum_{F \in \mathcal{F}} \left(\sum_{f \in F} \beta_f^F I_f \right).$$

Fix $F \in \mathcal{F}$ and consider a parametric functional $\phi(\beta)$ of the form

$$\phi(\beta) = \sum_{f \in F} p_f \beta_f^F. \tag{8.23}$$

We claim $\phi(\beta)$ *is estimable if and only if*

$$\begin{array}{c} \text{for each } G \in \mathcal{F} \text{ distinct from } F \text{ and each block } h \\ \text{of } H = F \wedge G, \text{ one has } \sum_{f \in F, f \subset h} p_f = 0, \end{array} \tag{8.24}$$

and then the GME of $\phi(\beta)$ is

$$\hat{\phi}(Y) = \sum_{f \in F} p_f \bar{Y}_f, \tag{8.25}$$

where $\bar{Y}_f$ is the average value of Y over f (see (2.3.55)).

First we show that $\phi(\beta)$ is nonestimable if condition (8.24) is false. Suppose that $\sum_{f \in F, f \subset h} p_f \neq 0$ for some block h of $H = F \wedge G$ for some $G \neq F$. h is a union of certain blocks of F, say $f_1, \ldots, f_I$, and also a union of certain blocks of G, say $g_1, \ldots, g_J$. Let β_1 be the element of B having all coordinates 0, and let β_2 be the element of B having all coordinates 0 except for $(\beta_2)_{f_1}^F = \cdots = (\beta_2)_{f_I}^F = 1$ and $(\beta_2)_{g_1}^G = \cdots = (\beta_2)_{g_J}^G = -1$. Then

$$X\beta_2 = \sum_{1 \leq i \leq I} I_{f_i} - \sum_{1 \leq j \leq J} I_{g_j} = I_h - I_h = 0 = X\beta_1$$

while

$$\phi(\beta_2) = \sum_{1 \leq i \leq I} p_{f_i} = \sum_{f \in F, f \subset h} p_f \neq 0 = \phi(\beta_1).$$

This violates condition (iii) of Proposition 8.4, so $\phi(\beta)$ is nonestimable.

Suppose next that condition (8.24) does hold. Put

$$v = \sum_{f \in F} \frac{p_f}{|f|} I_f,$$

where $|f|$ denotes the number of experimental units assigned to level f (see (2.3.29)). We claim that

$$E_\beta \langle v, Y \rangle = \phi(\beta)$$

for all $\beta \in B$. This will make $\phi(\beta)$ estimable and (8.25) will follow because $v \in M$ and $\langle v, Y \rangle = \sum_{f \in F} p_f \bar{Y}_f$ (see parts (i) of Propositions 8.4 and 8.16). Now

$$E_\beta \langle v, Y \rangle = \langle v, E_\beta Y \rangle = \sum_{G \in \mathcal{F}} \langle v, X_G \beta^G \rangle,$$

where

$$X_G \beta^G = \sum_{g \in G} \beta_g^G I_g.$$

Since

$$\langle v, X_F \beta^F \rangle = \sum_{f \in F} \sum_{f' \in F} \frac{p_f}{|f|} \beta_{f'}^F |f \cap f'| = \sum_{f \in F} p_f \beta_f^F = \phi(\beta),$$

it suffices to show $\langle v, X_G \beta^G \rangle = 0$ for $G \in \mathcal{F}$ distinct from F. Fix such a G and put $H = F \wedge G$. Because F and G are orthogonal factors, we have $|f \cap g| = |f| |g|/|h|$ whenever the blocks f of F and g of G are nested in the same block h of H (see

Proposition 2.3.57), while trivially $|f \cap g| = 0$ if f and g are nested in different blocks of H. Hence

$$\langle v, X_G \beta^G \rangle = \sum_{f \in F} \sum_{g \in G} \frac{p_f}{|f|} \beta_g^G |f \cap g|$$

$$= \sum_{h \in H} \sum_{f,g \subset H, f \in F, g \in G} \frac{p_f}{|f|} \beta_g^G \frac{|f||g|}{|h|}$$

$$= \sum_{h \in H} \frac{1}{|h|} \left[\left(\sum_{f \in F, f \subset h} p_f \right) \left(\sum_{g \in G, g \subset h} \beta_g^G |g| \right) \right] = 0,$$

the last equality holding by (8.24).   •

8.26 Exercise [3]. In the context of Example 8.21, show that for fixed $F \in \mathcal{F}$ and fixed $f_1, f_2 \in F$, the parametric functional $\phi(\beta) = \beta_{f_2}^F - \beta_{f_1}^F$ is estimable if and only if for each factor $G \in \mathcal{F}$ distinct from F, f_1 and f_2 are nested in the same block of $H = F \wedge G$.   ◇

8.27 Exercise [3]. In the context of Example 8.21, suppose that F and G are factors such that $\phi_F(\beta) = \sum_{f \in F} p_f^F \beta_f^F$ and $\phi_G(\beta) = \sum_{g \in G} p_f^G \beta_f^G$ are both estimable. Find the variances and covariances of the GMEs $\hat{\phi}_F(Y)$ and $\hat{\phi}_G(Y)$.   ◇

8.28 Exercise [3]. Consider the split-plot design of Exercise 2.3.66. Suppose $\mathcal{F} = \{P, A \times B\}$. For each choice of $F \in \mathcal{F}$, what parametric functionals $\phi(\beta) = \sum_{f \in F} p_f \beta_f^F$ are estimable, and what are their GMEs? Repeat for the case when $\mathcal{F}$ is augmented to $\mathcal{F}^* = \{P, A \times B, A, B, T\}$ by including those factors in $\mathcal{D}$ that lie below P and/or $A \times B$ in the factor structure diagram.   ◇

The exercises that follow deal with the case when certain linear restrictions are imposed on β. For the rest of this section suppose that R is a linear transformation mapping B into a inner product space C and modify (8.3) to read

> the random vector Y is weakly spherical in V
> and has mean $\mu = E(Y)$ of the form $X\beta$ for   (8.29)
> some unknown $\beta \in B$ such that $R\beta = 0$.

Put

$$M = \{ X\beta : \beta \in B \text{ and } R\beta = 0 \}. \tag{8.30}$$

8.31 Exercise [2]. Show that $\hat{\beta} \in B$ satisfies $X\hat{\beta} = P_M Y$ and $R\hat{\beta} = 0$ if and only if there exists a $c \in C$ such that $\hat{\beta}$ and c jointly satisfy the *normal equations*

$$X'X\hat{\beta} = X'Y + R'c \quad \text{and} \quad R\hat{\beta} = 0. \tag{8.32}$$

[Hint: Use $\mathcal{N}^{\perp}(R) = \mathcal{R}(R')$.]   ◇

8.33 Exercise [3]. Let ϕ be a parametric functional with coefficient vector $p \in B$, so that $\phi(\beta) = \langle p, \beta \rangle_B$ for all $\beta \in B$. Show that the following are equivalent:

  (i) $\phi(\beta)$ admits a linear unbiased estimator, in the sense that there exists a $v \in V$ such that $E_\beta \langle v, Y \rangle_V = \phi(\beta)$ for all $\beta \in B$ such that $R\beta = 0$,

(ii) there exists a linear functional ψ on M such that

$$\phi(\beta) = \psi(X\beta) \quad \text{for all } \beta \in B \text{ such that } R\beta = 0, \tag{8.34}$$

(iii) $\phi(\beta)$ is uniquely determined by $X\beta$ when $R\beta = 0$, in the sense that $\phi(\beta_1) = \phi(\beta_2)$ whenever $\beta_1, \beta_2 \in B$ satisfy $X\beta_1 = X\beta_2$ and $R\beta_1 = 0 = R\beta_2$,

(iv) $p \perp (\mathcal{N}(X) \cap \mathcal{N}(R))$,

(v) $p \in \mathcal{R}(X') + \mathcal{R}(R')$,

(vi) $p \in \mathcal{R}(X'X + RR')$.   ◇

8.35 Exercise [3]. Suppose ϕ satisfies the conditions of the previous exercise. Let ψ be the unique linear functional on M such that equation (8.34) holds. Show that $\hat{\phi}(Y) \equiv \hat{\psi}(Y)$ is the best linear unbiased estimator of $\phi(\beta)$ and that $\hat{\phi}(Y)$ can be written variously as:

(i) $\langle P_M v, Y \rangle_V$, for any $v \in V$ such that $E_\beta \langle v, Y \rangle_V = \phi(\beta)$ for all $\beta \in B$ with $R\beta = 0$,

(ii) $\langle P_M v, Y \rangle_V$ for any $v \in V$ and $c \in C$ such that $p = X'v + R'c$,

(iii) $\phi(\hat{\beta})$ for any $\hat{\beta}$ satisfying the normal equations (8.32), or

(iv) $\langle b, X'Y + R'c \rangle_B$ for any $b \in B$ such that $p = (X'X + RR')b$ and any $c \in C$ satisfying the normal equations (8.32).   ◇

9. Problem set: Quantifying the Gauss-Markov theorem

Let V, $\langle \cdot, \cdot \rangle$, Y, and M play their usual roles in the GLM — in particular, Y is weakly spherical with respect to $\langle \cdot, \cdot \rangle$ and M is the subspace of possible means of Y. Let $\langle \cdot, \cdot \rangle^* = \langle \cdot, \Delta \cdot \rangle$ be another inner product on V. In this problem set you will work out an explicit, albeit somewhat cumbersome, expression for the ratio

$$\frac{\text{Var}\big(\psi(P_M^* Y)\big)}{\text{Var}\big(\psi(P_M Y)\big)}$$

of the variance of the false GME of the linear functional $\psi(\mu)$ to the variance of the true GME of $\psi(\mu)$. The ratio depends on the extent to which vectors $\perp^*$ to M fail to be $\perp$ to M, so the problem set begins with some preliminary considerations on measuring the nonorthogonality of subspaces.

Let K and L be two subspaces of V; assume $K \neq 0$ and $L \neq 0$. Set

$$\theta(K, L) = \sup\{ \langle x, y \rangle : x \in K, \ y \in L, \ \|x\| = 1 = \|y\| \}. \tag{9.1}$$

Topological considerations show that the sup is attained.

A. Show that

$$0 \leq \theta(K, L) \leq 1 \tag{9.2}$$

with

$$\theta(K, L) = 0 \iff K \perp L \tag{9.3}$$

$$\theta(K, L) = 1 \iff K \cap L \neq 0.$$

[Hint: Use the Cauchy-Schwarz inequality (2.1.21).]   ○

B. Suppose $x \in K$ and $y \in L$ with $\theta(K, L) = \langle x, y \rangle$ and $\|x\| = 1 = \|y\|$. Put

$$K' = \text{orthogonal complement of } [x] \text{ in } K = K - [x]$$
$$L' = \text{orthogonal complement of } [y] \text{ in } L = L - [y].$$

Show that

$$x \perp L' \quad \text{and} \quad y \perp K'.$$

[Hint: Were there a nonzero $x' \in K'$ with $\langle x', y \rangle > 0$, then $\langle (x + \delta x')/\|x + \delta x'\|, y \rangle$ would exceed $\langle x, y \rangle$ for small positive δ.]   ○

C. Show that there exist orthonormal bases $x_1, \ldots, x_k$ for K and $y_1, \ldots, y_\ell$ for L such that

$$\langle x_i, y_j \rangle = 0 \quad \text{for } i \neq j, \ 1 \leq i \leq k, \ 1 \leq j \leq \ell,$$

$$\langle x_g, y_g \rangle = \theta(K - [x_1, \ldots, x_{g-1}], L - [y_1, \ldots, y_{g-1}]), \tag{9.4}$$
$$\text{for } 1 \leq g \leq h \equiv \min(k, \ell).$$

The bases $x_1, \ldots, x_k$ and $y_1, \ldots, y_\ell$ are called *canonical bases* for K and L. [Hint: Use **B** repeatedly.]   ○

D. Let x and y be given in V with $\|x\| = 1 = \|y\|$ and $\langle x, y \rangle \geq 0$. Show that the linear transformation mapping V into V that sends x into itself and that annihilates y and $[x, y]^\perp$ is

$$\frac{1}{1 - \theta^2} \left(\langle \cdot, x \rangle - \theta \langle \cdot, y \rangle \right) x$$

with

$$\theta = \langle x, y \rangle = \theta([x], [y]).$$ ○

That finishes the preliminaries. Now let $M^{\perp^*}$ denote the orthogonal complement of M with respect to $\langle \cdot, \cdot \rangle^*$. Let $x_1, \ldots, x_k$ and $y_1, \ldots, y_\ell$ be canonical bases for M and $M^{\perp^*}$, respectively. Put

$$h = \min(k, \ell)$$

and set

$$\theta_g = \theta(M - [x_1, \ldots, x_{g-1}], M^{\perp^*} - [y_1, \ldots, y_{g-1}]) \quad \text{for } 1 \leq g \leq h.$$

Note $1 > \theta_1 \geq \theta_2 \cdots \geq \theta_h \geq 0$ by **A**.

E. Show that for each $v \in V$,

$$P_M v = \sum_{1 \leq i \leq k} \langle v, x_i \rangle x_i$$

whereas

$$P_M^* v = \sum_{1 \leq g \leq h} \frac{1}{1 - \theta_g^2} \left(\langle v, x_g \rangle - \theta_g \langle v, y_g \rangle \right) x_g + \sum_{h+1 \leq i \leq k} \langle v, x_i \rangle x_i. \tag{9.5}$$

Deduce that the difference between the false GME $\hat{\mu}^* = P_M^* Y$ and the true GME $\hat{\mu} = P_M Y$ is given by

$$\hat{\mu}^* - \hat{\mu} = \sum_{1 \leq g \leq h} \frac{\theta_g}{1 - \theta_g^2} \langle Y, \theta_g x_g - y_g \rangle x_g. \tag{9.6}$$

[Hint: For (9.5), show that the right-hand side is x_i when $v = x_i$, for $1 \leq i \leq k$, and is 0 when $v = y_j$, for $1 \leq j \leq \ell$.]   ○

F. Show that the linear functionals

$$f_g(Y) = \frac{\theta_g}{1 - \theta_g^2} \langle Y, \theta_g x_g - y_g \rangle, \quad 1 \leq g \leq h,$$

appearing in (9.6) are uncorrelated among themselves and with the functionals

$$\langle Y, x_i \rangle, \quad 1 \leq i \leq k,$$

and that the $f_g(Y)$'s have variances

$$\mathrm{Var}\big(f_g(Y)\big) = \sigma^2 \frac{\theta_g^2}{1 - \theta_g^2}.$$

[Hint: Compute.]   ○

G. Now let $\psi(\mu)$ be a linear functional of μ. Let $c \in M$ be the coefficient vector of ψ with respect to $\langle \cdot, \cdot \rangle$:

$$\psi(m) = \langle c, m \rangle \quad \text{for } m \in M.$$

Put

$$\hat{\psi} = \psi(P_M Y), \quad \hat{\psi}^* = \psi(P_M^* Y).$$

Show that

$$\frac{\mathrm{Var}(\hat{\psi}^*)}{\mathrm{Var}(\hat{\psi})} = 1 + \frac{\sum_{1 \leq g \leq h} c_g^2 \frac{\theta_g^2}{1 - \theta_g^2}}{\sum_{1 \leq g \leq k} c_g^2}, \tag{9.7}$$

where $c_g = \langle c, x_g \rangle$ for $1 \leq g \leq k$.
[Hint: Use (9.6) and **F**.]   ○

H. Let τ be the ratio of the largest to the smallest eigenvalues of the operator Δ figuring in the definition of $\langle \cdot, \cdot \rangle^*$. Show that

$$\frac{\mathrm{Var}(\hat{\psi}^*)}{\mathrm{Var}(\hat{\psi})} \leq \frac{(1 + \tau)^2}{4\tau}. \tag{9.8}$$

[Hint: $\mathrm{Var}(\hat{\psi}^*)/\mathrm{Var}(\hat{\psi}) \leq 1/(1 - \theta_1^2)$; now use Cleveland's identity (2.6.1).]   ○

I. Exhibit an M and a ψ for which equality holds in (9.8).
[Hint: First show that $(1 + \tau)^2/(4\tau) = 1/(1 - \theta_1^2)$.]   ○

J. Suppose $V = \mathbb{R}^n$, $\langle \cdot, \cdot \rangle^*$ is the dot-product and $Y = (Y_1, \ldots, Y_n)^T$ with

$$\mathrm{Var}(Y_i) = \sigma^2 \quad \text{for each } i$$

and

$$\mathrm{Correlation}(Y_i, Y_j) = \rho \quad \text{for } i \neq j.$$

Show that the bound on the right-hand side of (9.8) is

$$\frac{\big(2 + (n-2)\rho\big)^2}{4(1-\rho)\big(1 + (n-1)\rho\big)} . \tag{9.9} \circ$$

CHAPTER 5

NORMAL THEORY: ESTIMATION

Let V, $\langle \cdot, \cdot \rangle$, M, and Y have their customary meanings in the GLM with the weak sphericity setup. *Throughout this chapter* we suppose now in addition that Y is normally distributed in V:

$$Y \sim N_V(\mu, \sigma^2 I_V) \quad \text{with } \mu \in M \text{ and } \sigma^2 > 0. \tag{0.1}$$

We show that as an estimator of μ, $P_M Y$ has various nice properties, to wit: (i) It is the maximum likelihood estimator of μ; (ii) it has minimum dispersion in the class of all (linear and nonlinear) unbiased estimators of μ; and (iii) it is minimax with respect to mean square error. On the debit side, we exhibit the James-Stein phenomenon: $P_M Y$ is inadmissible relative to mean square error when $\dim(M) \geq 3$. Some admissible minimax estimators are developed in the problem set.

1. Maximum likelihood estimation

Relative to Lebesgue measure on V, Y has density (see (3.7.6))

$$
\begin{aligned}
f_{\mu,\sigma^2}(y) &= \frac{1}{(2\pi)^{n/2}|\Sigma|^{1/2}} \, e^{-\frac{1}{2}\langle y-\mu, \Sigma^{-1}(y-\mu)\rangle} \\
&= \frac{1}{(2\pi\sigma^2)^{n/2}} \, e^{-\frac{1}{2}\|y-\mu\|^2/\sigma^2} \\
&= \frac{1}{(2\pi\sigma^2)^{n/2}} \, e^{-\frac{1}{2}\|P_M y-\mu\|^2/\sigma^2} \, e^{-\frac{1}{2}\|Q_M y\|^2/\sigma^2},
\end{aligned}
\tag{1.1}
$$

where $n = \dim(V)$ and $\Sigma = \Sigma(Y)$. The *maximum likelihood estimators* $\hat{\mu}_{\mathrm{MLE}}$ and $\hat{\sigma}^2_{\mathrm{MLE}}$ of μ and σ^2 are the values of μ and σ^2 that maximize $f_{\mu,\sigma^2}(Y)$. Now, regardless of the value of σ, the maximum of $f_{\mu,\sigma^2}(Y)$ with respect to μ occurs at

$$\hat{\mu}_{\mathrm{MLE}} = P_M Y; \tag{1.2}$$

$\hat{\sigma}^2_{\text{MLE}}$ is therefore the value of σ^2 maximizing

$$\frac{1}{(\sigma^2)^{n/2}} \, e^{-\frac{1}{2}\|Q_M Y\|^2/\sigma^2} \; .$$

Since

$$\frac{d}{d\theta} \log\left(\frac{1}{\theta^{n/2}} \, e^{-\frac{1}{2} a/\theta}\right) = \frac{d}{d\theta}\left(-\frac{n}{2}\log(\theta) - \frac{a}{2\theta}\right) = \frac{1}{2\theta}\left(-n + \frac{a}{\theta}\right)$$

is strictly positive for $0 < \theta < a/n$ and strictly negative for $a/n < \theta$, we have

$$\hat{\sigma}^2_{\text{MLE}} = \frac{\|Q_M Y\|^2}{n} \; . \tag{1.3}$$

The joint distribution of $\hat{\mu}_{\text{MLE}}$ and $\hat{\sigma}^2_{\text{MLE}}$ is readily obtained from our basic distributional results under normality (see Theorem 3.8.2):

$$\hat{\mu}_{\text{MLE}} \sim N_M(\mu, \sigma^2 I_M) \tag{1.4}$$

independently of

$$\hat{\sigma}^2_{\text{MLE}} \sim \sigma^2 \chi^2_{d(M^\perp)}/n. \tag{1.5}$$

1.6 Exercise [2]. Suppose $\psi(\mu)$ is a linear functional on M, $\hat{\psi}(Y)$ is its GME, and $\hat{\sigma}_{\hat{\psi}} = \hat{\sigma}\|\psi\|$ is its estimated standard error (see (4.4.2)). Show that

$$\frac{\hat{\psi}(Y) - \psi(\mu)}{\hat{\sigma}_{\hat{\psi}}}$$

has a t distribution with $d(M^\perp)$ degrees of freedom.   ◇

2. Minimum variance unbiased estimation

Set

$$\hat{\mu} = \hat{\mu}_{\text{MLE}} = P_M Y = \text{GME of } \mu \tag{2.1}$$

$$\hat{\sigma}^2 = \frac{n}{d(M^\perp)} \, \hat{\sigma}^2_{\text{MLE}} = \frac{\|Q_M Y\|^2}{d(M^\perp)} \; . \tag{2.2}$$

Evidently

$$E_{\mu,\sigma^2}(\hat{\mu}) = \mu$$

$$E_{\mu,\sigma^2}(\hat{\sigma}^2) = \sigma^2$$

for all $\mu \in M$ and $\sigma^2 > 0$. Thus $\hat{\mu}$ and $\hat{\sigma}^2$ are unbiased estimators of μ and σ^2, respectively. According to the Lehmann-Scheffé theorem, they have minimum dispersion in the class of *all* unbiased estimators because, as we will now show, they are functions of a complete sufficient statistic. Recall that a statistic $T(Y)$ is: (i) *sufficient* for μ and σ^2 if for each possible value t of T, the conditional distribution of

Y given $T(Y) = t$ does not depend on the parameters μ, σ^2; and (ii) *complete* if whenever g is a function such that

$$E_{\mu,\sigma^2} g(T(Y)) = 0 \quad \text{for all } \mu \text{ and } \sigma^2,$$

then

$$P_{\mu,\sigma^2}\big(g(T(Y)) \neq 0\big) = 0 \quad \text{for all } \mu \text{ and } \sigma^2.$$

The Lehmann-Scheffé theorem (see Bickel and Doksum (1977, p. 122)) states that when $T(Y)$ is both sufficient and complete, each function of $T(Y)$ is the minimum dispersion unbiased estimator of its expected value.

To find a complete sufficient statistic, write the density of Y as

$$
\begin{aligned}
f_{\mu,\sigma^2}(y) &= \frac{1}{(2\pi\sigma^2)^{n/2}} \, e^{-\frac{1}{2}\|y-\mu\|^2/\sigma^2} \\
&= \frac{1}{(2\pi\sigma^2)^{n/2}} \, e^{-\frac{1}{2}\|y\|^2/\sigma^2} \, e^{\langle y, \mu/\sigma^2\rangle} \, e^{-\frac{1}{2}\|\mu\|^2/\sigma^2} \\
&= \frac{1}{(2\pi\sigma^2)^{n/2}} \, e^{-\frac{1}{2}\|y\|^2/\sigma^2} \, e^{\langle P_M y, \mu/\sigma^2\rangle} \, e^{-\frac{1}{2}\|\mu\|^2/\sigma^2} \\
&= C(\theta_1, \ldots, \theta_p, \theta_{p+1}) \, e^{\sum_{1\leq i\leq p+1} T_i(y)\theta_i}, \quad\quad (2.3)
\end{aligned}
$$

where, with $b_1, \ldots, b_p$ denoting an orthonormal basis for M,

$$T_i(y) = \langle P_M y, b_i\rangle, \qquad \theta_i = \langle \mu/\sigma^2, b_i\rangle, \quad \text{for } i = 1, \ldots, p,$$

$$T_{p+1}(y) = \|y\|^2, \qquad \theta_{p+1} = -\frac{1}{2\sigma^2},$$

and

$$C(\theta_1, \ldots, \theta_p, \theta_{p+1}) = \frac{1}{\pi^{n/2}} \, (-\theta_{p+1})^{n/2} \, e^{-\frac{1}{2}\sum_{1\leq i\leq p}\theta_i^2}.$$

Notice that as (μ, σ^2) ranges over $M \times (0, \infty)$, $\theta \equiv (\theta_1, \ldots, \theta_p, \theta_{p+1})$ ranges over

$$\Theta = \mathbb{R}^p \times (-\infty, 0).$$

It follows from (2.3) and the factorization criterion for sufficiency (see Lehmann (1959, p. 49)) that

$$T(Y) = \big(T_1(Y), \ldots, T_p(Y), T_{p+1}(Y)\big)$$

is sufficient; moreover, $T(Y)$ is complete because the possible distributions of $T(Y)$ constitute an exponential family and Θ has a nonempty interior as a subset of $\mathbb{R}^{p+1}$ (see Lehmann (1959, p. 132)). To finish off the argument, note that

$$\hat{\mu} = P_M Y = \sum_{1\leq i\leq p} T_i(Y) b_i$$

$$\hat{\sigma}^2 = \frac{\|Q_M Y\|^2}{d(M^\perp)} = \frac{\|Y\|^2 - \|P_M Y\|^2}{d(M^\perp)} = \frac{T_{p+1}(Y) - \sum_{1\leq i\leq p} T_i^2(Y)}{d(M^\perp)}$$

are indeed functions of $T(Y)$.

2.4 Exercise [4]. Suppose $g: V \to \mathbb{R}$ is a function such that

$$J(\mu, \sigma^2) \equiv E_{\mu, \sigma^2}(g(Y))$$

is defined and finite for all $\mu \in M$ and all $\sigma^2 > 0$. Show that J has continuous partial derivatives with respect to μ and σ^2 of all orders (indeed, is analytic). [Hint: See Lehmann (1959, p. 52).]   ◇

3. Minimaxity of $P_M Y$

For the rest of this chapter, with exceptions as noted, suppose σ^2 is known; for simplicity take $\sigma^2 = 1$, so

$$Y \sim N_V(\mu, I_V) \quad \text{with } \mu \in M. \tag{3.1}$$

We will investigate properties of the GME $P_M Y$ from a decision-theoretic viewpoint relative to *mean square error*, under which the *risk* of using an estimator $\hat{\mu}$ when μ obtains is

$$R(\hat{\mu}; \mu) \equiv E_\mu \|\hat{\mu} - \mu\|^2 = \text{trace}(\Sigma(\hat{\mu})) + \|E_\mu \hat{\mu} - \mu\|^2$$

(see (4.7.3)). Because $P_M Y$ has mean μ and dispersion operator $P_M P'_M = P_M$ in V, the risk of $P_M Y$ is

$$R(P_M Y; \mu) = E_\mu \|P_M Y - \mu\|^2 = \text{trace}(P_M) + \|0\|^2 = \dim(M);$$

note that $P_M Y$ has constant risk.

Recall that an estimator $\hat{\mu} = \hat{\mu}(Y)$ of μ is said to be *minimax* if

$$\bar{R}(\hat{\mu}) \equiv \sup_{\mu \in M} R(\hat{\mu}; \mu)$$
$$= \inf_\delta \bar{R}(\delta) \equiv \text{value of the statistical game},$$

the infimum being taken over all estimators $\delta = \delta(Y)$ of μ. $\hat{\mu}$ is said to be *Bayes versus a prior* Π on μ if

$$R(\hat{\mu}; \Pi) \equiv \int_M R(\hat{\mu}; \mu)\, \Pi(d\mu) = \inf_\delta R(\delta; \Pi) \equiv \mathcal{B}(\Pi). \tag{3.2}$$

$\hat{\mu}$ is *ϵ-Bayes versus* Π if $R(\hat{\mu}; \Pi) - \mathcal{B}(\Pi) \leq \epsilon$. $\hat{\mu}$ is *extended Bayes* if for each n there exists a prior Π_n such that $\hat{\mu}$ is $1/n$-Bayes versus Π_n. We will show that $P_M Y$ is minimax using the following lemma.

3.3 Lemma. *Suppose $\hat{\mu}$ is an estimator of μ having finite constant risk, say r. If there exists a sequence $\{\Pi_n\}$ of priors on μ such that $\mathcal{B}(\Pi_n) \to r$, then $\hat{\mu}$ is extended Bayes and minimax, and r is the value of the statistical game.*

Proof. $\hat{\mu}$ is extended Bayes because $R(\hat{\mu}; \Pi_n) = r$ for each n. To see that $\hat{\mu}$ is minimax, note that for any estimator δ and any n,

$$\bar{R}(\delta) = \sup_{\mu} R(\delta; \mu) \geq \int R(\delta; \mu) \, \Pi_n(d\mu) \geq \mathcal{B}(\Pi_n);$$

letting $n \to \infty$ gives

$$\bar{R}(\delta) \geq r = \bar{R}(\hat{\mu}). \qquad \blacksquare$$

We now need to work out a recipe for a Bayes rule versus a prior Π on μ. For this, let Π be given and, for notational convenience, let Θ denote an M-valued random vector defined on the same probability space as Y in such a way that the marginal distribution of Θ is Π and, for each $\mu \in M$, the conditional distribution of Y, given $\Theta = \mu$, is $N_V(\mu, I_V)$.

For any estimator δ,

$$\begin{aligned}
R(\delta; \Pi) &= \int_M E_\mu \left(\|\delta(Y) - \mu\|^2 \right) \Pi(d\mu) \\
&= \int_M E\left(\|\delta(Y) - \Theta\|^2 \mid \Theta = \mu \right) \Pi(d\mu) \\
&= E\left(\|\delta(Y) - \Theta\|^2 \right) \\
&= \int_V E\left(\|\delta(Y) - \Theta\|^2 \mid Y = y \right) P(dy) \\
&= \int_V E_y \left(\|\Theta - \delta(y)\|^2 \right) P(dy),
\end{aligned}$$

where P denotes the marginal law of Y, the symbol $\mid$ is to be read as "given," and E_y denotes conditional expectation, given $Y = y$. The idea now is to set

$$\rho(y) = E_y(\Theta)$$

and to write

$$\begin{aligned}
E_y \left(\|\Theta - \delta(y)\|^2 \right) &= E_y \left(\|(\Theta - \rho(y)) + (\rho(y) - \delta(y))\|^2 \right) \\
&= E_y \left(\|\Theta - \rho(y)\|^2 + 2\langle \Theta - \rho(y), \rho(y) - \delta(y) \rangle \right. \\
&\quad \left. + \|\rho(y) - \delta(y)\|^2 \right) \\
&= E_y \left(\|\Theta - \rho(y)\|^2 \right) + \|\rho(y) - \delta(y)\|^2,
\end{aligned} \qquad (3.4)$$

whence

$$R(\delta; \Pi) \geq R(\rho; \Pi).$$

Modulo the nuisance of checking that $\rho(y)$ is well defined (that is, that Θ actually has an E_y-expectation) and that E_y distributes over the summations in (3.4), this proves

3.5 Lemma. *For any prior Π on μ, the Bayes estimator versus Π is $\rho(Y)$, where*

$$\rho(y) = E(\Theta \mid Y = y) \equiv E_y(\Theta) \tag{3.6}$$

is the mean of the posterior distribution of μ given $Y = y$ and where

$$\mathcal{B}(\Pi) = R(\rho; \Pi) = \int R(\rho; \mu) \, \Pi(d\mu). \tag{3.7}$$

3.8 Exercise [3]. Show that the fact that $\mathcal{B}(\Pi) < \infty$ ($P_M Y$ has bounded risk) implies that

$$E_y(\|\Theta\|^2) < \infty$$

for P-almost all y, and thus that the lemma is valid (measure-theoretic considerations set aside).
[Hint: If $E_y(\|\Theta + v\|^2)$ is finite for some $v \in V$, then it is finite for all $v \in V$, in particular for $v = 0$.] ⬦

A natural activity now is to search for priors Π such that the posterior means can be calculated without undue difficulty. It turns out that certain normal priors meet this desideratum, thanks to the following result.

3.9 Lemma. *Let W_i, $\langle \cdot, \cdot \rangle_{W_i}$, $i = 1, 2$, be two inner product spaces and for each i, let Z_i be a W_i-valued random vector. Let*

$$W = W_1 \times W_2$$

be endowed with the inner product

$$\langle (v_1, v_2), (w_1, w_2) \rangle_W = \langle v_1, w_1 \rangle_{W_1} + \langle v_2, w_2 \rangle_{W_2}$$

and let Z be the W-valued random vector

$$(Z_1, Z_2).$$

Put

$$\mu_i = E(Z_i), \qquad\qquad i = 1, 2,$$
$$\Sigma_{ij} = \mathrm{Cov}(Z_i, Z_j), \qquad i, j = 1, 2,$$

and assume that Σ_{22} is nonsingular. The following are then equivalent:

(i) *Z has a normal distribution in W;*

(ii) *Z_2 has a normal distribution in W_2 and the conditional distributions of Z_1, given $Z_2 = z_2$, are normal in W_1 with conditional means depending affinely on z_2 and with constant conditional dispersions, that is,*

$$\mathcal{L}(Z_1 \mid Z_2 = z_2) = N_{W_1}(w_1 + Az_2, B) \quad \text{for all } z_2 \in W_2$$

for some $w_1 \in W_1$ and some linear transformations $A: W_2 \to W_1$ and $B: W_1 \to W_1$;

(iii) $Z_2 \sim N_{W_2}(\mu_2, \Sigma_{22})$ and

$$\mathcal{L}(Z_1 \mid Z_2 = z_2) = N_{W_1}\left(\mu_1 + \Sigma_{12}\Sigma_{22}^{-1}(z_2 - \mu_2),\ \Sigma_{11} - \Sigma_{12}\Sigma_{22}^{-1}\Sigma_{21}\right)$$

for each $z_2 \in W_2$.

3.10 Exercise [4]. Prove the lemma. To show that (ii) implies (i), argue that for any $c_1 \in W_1$ and $c_2 \in W_2$, the linear functional

$$\langle (c_1, c_2), (Z_1, Z_2) \rangle_W = \langle c_1, Z_1 \rangle_{W_1} + \langle c_2, Z_2 \rangle_{W_2}$$

of Z is normally distributed, by showing it has characteristic function

$$E\left(e^{i\xi(\langle c_1, Z_1 \rangle + \langle c_2, Z_2 \rangle)}\right)$$

($\xi \in \mathbb{R}$) of the form $e^{i\#_1\xi - \#_2\xi^2/2}$ for some constants $\#_1, \#_2 \in \mathbb{R}$. To show that (i) implies (iii), start with the observation that $Z_1 - \Sigma_{12}\Sigma_{22}^{-1}Z_2$ and Z_2 are independent.   ◇

3.11 Example. Suppose the M-valued random vector Θ has marginal distribution

$$\mathcal{L}(\Theta) = N_M(0, \lambda I_M)$$

for some $\lambda > 0$ and for each $\mu \in M$, the conditional distribution of Y given $\Theta = \mu$ is

$$\mathcal{L}(Y \mid \Theta = \mu) = N_V(\mu, I_V),$$

as in (3.1). Part (ii) of the preceding lemma tells us that Y and Θ are jointly normally distributed, and by part (iii) of the lemma we know that

(a) $E(\Theta) = 0$,

(b) $\Sigma_{\Theta\Theta} = \lambda I_M$,

(c) $E(Y) + \Sigma_{Y\Theta}\Sigma_{\Theta\Theta}^{-1}(\mu - E\Theta) = \mu$ for all $\mu \in M$,

(d) $\Sigma_{YY} - \Sigma_{Y\Theta}\Sigma_{\Theta\Theta}^{-1}\Sigma_{\Theta Y} = I_V$,

where, for example, $\Sigma_{\Theta\Theta} = \mathrm{Cov}(\Theta, \Theta)$. Let P_M° be P_M considered as a linear transformation from V onto M, rather than as a linear transformation from V into V. (P_M° and P_M are alike in that $P_M^\circ v = P_M v$ for all $v \in V$ but are different in that their adjoints have different domains.) Utilizing (a)–(d) and letting $i_M : M \to V$ be defined by $i_M(\mu) = \mu$, we get

$$E(Y) = 0 \qquad \text{(put } \mu = 0 \text{ in (c))}$$

$$\Sigma_{Y\Theta} = \lambda i_M \qquad \text{((c) now reads } \Sigma_{Y\Theta}\mu/\lambda = \mu)$$

$$\Sigma_{\Theta Y} = \lambda P_M^\circ \qquad (\Sigma_{\Theta Y} = \Sigma_{Y\Theta}' \text{ and } i_M' = P_M^\circ)$$

$$\Sigma_{YY} = I_V + \lambda i_M P_M^\circ \qquad \text{(by (d))}$$

$$= I_V + \lambda P_M = (1 + \lambda)P_M + Q_M$$

$$\Sigma_{YY}^{-1} = \frac{1}{1+\lambda} P_M + Q_M$$

$$\Sigma_{\Theta Y} \Sigma_{YY}^{-1} = \frac{\lambda}{1+\lambda} P_M^{\circ} P_M + \lambda P_M^{\circ} Q_M = \frac{\lambda}{1+\lambda} P_M^{\circ}$$

$$\Sigma_{\Theta\Theta} - \Sigma_{\Theta Y}\Sigma_{YY}^{-1}\Sigma_{Y\Theta} = \lambda I_M - \frac{\lambda^2}{(1+\lambda)} I_M = \frac{\lambda}{1+\lambda} I_M.$$

It follows, again by part (iii) of the lemma, that the marginal distribution of Y is

$$\mathcal{L}(Y) = N_V(0, I_V + \lambda P_M),$$

and, for each $y \in V$, the conditional distribution of Θ given $Y = y$ is

$$\mathcal{L}(\Theta \mid Y = y) = N_M\left(\frac{\lambda}{1+\lambda} P_M y, \; \frac{\lambda}{1+\lambda} I_M\right). \qquad \bullet$$

We now know that

$$\delta_\lambda(Y) \equiv \frac{\lambda}{1+\lambda} P_M Y = \frac{\lambda}{1+\lambda} P_M Y + \frac{1}{1+\lambda} 0 \qquad (3.12)$$

is Bayes versus the prior

$$\Pi_\lambda = N_M(0, \lambda I_M).$$

Note that $\delta_\lambda(Y)$ is a convex combination of the prior mean $E(\Theta) = 0$ and the GME $\hat{\mu} = P_M Y$ of μ, with respective weights $\frac{1}{1+\lambda}$ and $\frac{\lambda}{1+\lambda}$ depending on the diffuseness λ of the prior. The risk of δ_λ is easily computed:

$$\begin{aligned}
R\left(\frac{\lambda}{1+\lambda} P_M Y; \mu\right) &= E_\mu\left(\left\|\frac{\lambda}{1+\lambda} P_M Y - \mu\right\|^2\right) \\
&= E_\mu\left(\left\|\frac{\lambda}{1+\lambda}(P_M Y - \mu) - \frac{1}{1+\lambda}\mu\right\|^2\right) \\
&= \left(\frac{\lambda}{1+\lambda}\right)^2 E_\mu(\|P_M Y - \mu\|^2) \\
&\quad - 2\frac{\lambda}{1+\lambda}\frac{1}{1+\lambda} E_\mu\langle P_M Y - \mu, \mu\rangle + E_\mu\left\|\frac{\mu}{1+\lambda}\right\|^2 \\
&= \left(\frac{\lambda}{1+\lambda}\right)^2 \dim(M) + \frac{1}{(1+\lambda)^2}\|\mu\|^2 \\
&= \left(\frac{\lambda}{1+\lambda}\right)^2 R(P_M Y; \mu) + \frac{1}{(1+\lambda)^2} R(0; \mu).
\end{aligned}$$

Since $\Theta \sim \Pi_\lambda$ has a weakly spherical distribution in M with mean 0, we have

$$E\|\Theta\|^2 = \lambda \dim(M) + \|E(\Theta)\|^2 = \lambda \dim(M),$$

so

$$\mathcal{B}(\Pi_\lambda) = R(\delta_\lambda; \Pi_\lambda) = E\big(R(\delta_\lambda; \Theta)\big)$$
$$= \Big(\frac{\lambda}{1+\lambda}\Big)^2 \dim(M) + \frac{\lambda}{(1+\lambda)^2}\dim(M)$$
$$= \Big(\frac{\lambda}{1+\lambda}\Big)\dim(M). \tag{3.13}$$

Note that as $\lambda \to \infty$,

$$\mathcal{B}(\Pi_\lambda) \to \dim(M) = R(P_M Y; \cdot).$$

By Lemma 3.3, the constant risk estimator $P_M Y$ is extended Bayes and minimax.

3.14 Exercise [3]. Show that when $\sigma^2 > 0$ is unknown, $P_M Y$ is minimax with respect to *normalized mean square error*

$$R(\hat{\mu}; \mu, \sigma^2) \equiv E_{\mu,\sigma^2}\Big(\frac{\|\hat{\mu} - \mu\|}{\sigma}\Big)^2. \tag{3.15}$$

[Hint: This is easily deduced from the minimaxity of $P_M Y$ when $\sigma^2 = 1$.]   ◇

4. James-Stein estimation

From a decision-theoretic viewpoint, it would be ideal if $P_M Y$ were not only minimax — providing the best protection against the worst possible risk — but also admissible — unable to be improved upon. Recall that for an estimator δ of μ to be *admissible*, there can be no other estimator δ^* such that

$$R(\delta^*; \mu) \le R(\delta; \mu) \quad \text{for all } \mu \in M,$$
$$R(\delta^*; \mu) < R(\delta; \mu) \quad \text{for some } \mu \in M. \tag{4.1}$$

It turns out that $P_M Y$ is admissible if $\dim(M) \le 2$. Surprisingly, however, $P_M Y$ is inadmissible for $\dim(M) \ge 3$, as we will now show by exhibiting an estimator with an everywhere smaller risk.

Consider the Bayesian setup of the previous section, where

$$\mu \sim N_M(0, \lambda I_M)$$

$$\mathcal{L}(Y \mid \mu) = N_V(\mu, I_V).$$

We saw that

$$\frac{\lambda}{1+\lambda} P_M Y = \Big(1 - \frac{1}{1+\lambda}\Big) P_M Y \tag{4.2}$$

was the Bayes rule in this case. This rule is admissible, but a potential drawback to its use is that we do not know λ. We can, however, estimate λ or, more to the point, $1/(1+\lambda)$ from the data Y, as follows. Put

$$X = P_M Y, \tag{4.3}$$

considered as an M-valued random vector. By Example 3.11, the (marginal) distribution of Y is $N_V(0, I_V + \lambda P_M)$, so $X \sim N_M(0, (1 + \lambda)I_M)$ and

$$S \equiv \|X\|^2 \sim (1 + \lambda)\chi_p^2, \tag{4.4}$$

where

$$p = \dim(M). \tag{4.5}$$

Now the first reciprocal moment of χ_p^2 is

$$\int_0^\infty \frac{1}{t} \left((\text{density of } \chi_p^2)(t)\right) dt = \int_0^\infty \frac{1}{t} \frac{1}{\Gamma(p/2)2^{p/2}} t^{p/2-1} e^{-t/2} dt$$

$$= \begin{cases} \dfrac{\Gamma\left(\frac{p-2}{2}\right) 2^{\frac{p-2}{2}}}{\Gamma(p/2)2^{p/2}} = \dfrac{1}{(p/2 - 1)2} = \dfrac{1}{p - 2}, & \text{if } p \geq 3, \\ \infty, & \text{if } p \leq 2. \end{cases} \tag{4.6}$$

Thus for $p \geq 3$, $\hat{B}(S) \equiv (p-2)/S$ unbiasedly estimates $B(\lambda) \equiv 1/(1+\lambda)$ and by (4.2), the estimator

$$(1 - \hat{B}(S))X = \left(1 - \frac{p-2}{S}\right)X$$

is worthy of consideration. A little more generally, we may consider rules of the form

$$\left(1 - \frac{c}{S}\right)X \tag{4.7}$$

for arbitrary constants c. Like the Bayes rules from which they were extrapolated, such estimates may be expected to have small risk for μ near 0; on the other hand, for large μ such rules will with high probability agree closely with X and so should have risk close to that of X.

There is another illuminating argument leading to rules of the form (4.7). A classical statistician might well approach the problem of estimating μ from X by fitting a simple model such as $\mu = \beta X + error$ with $\beta \in \mathbb{R}$. The choice of β minimizing $\|error\|^2$ is

$$\hat{\beta} = \frac{\langle X, \mu \rangle}{\|X\|^2} = \frac{\langle X, \mu \rangle}{S}.$$

One can not use $\hat{\beta}X$ as an estimator because $\langle X, \mu \rangle$ depends on μ, which is unknown. However, the expected value of $\langle X, \mu \rangle$, namely $\|\mu\|^2$, is unbiasedly estimated by $\|X\|^2 - p = S - p$. Replacing the unobservable $\hat{\beta}$ by the observable $(1 - p/S)$ leads to (4.7) with $c = p$.

4.8 Theorem (James-Stein). *Suppose $p \equiv \dim(M) \geq 3$. The estimator of μ of the form (4.7) having minimum risk, uniformly in μ, is*

$$\hat{\mu}_{JS} \equiv \left(1 - \frac{p-2}{S}\right)X = \left(1 - \frac{p-2}{\|P_M Y\|^2}\right)P_M Y; \tag{4.9}$$

its risk is

$$R(\hat{\mu}_{JS}; \mu) = E_\mu \|\hat{\mu}_{JS} - \mu\|^2$$
$$= p - (p-2)^2 E_\mu\left(\frac{1}{S}\right) = p - (p-2)^2 \mathcal{E}_p(\|\mu\|^2), \tag{4.10}$$

where, for $t \geq 0$,

$$\mathcal{E}_p(t) \equiv \sum_{0 \leq k < \infty} e^{-t/2} \frac{(t/2)^k}{k!} \frac{1}{p-2+2k} = E\left(\frac{1}{p-2+2K}\right), \tag{4.11}$$

K being a Poisson random variable with mean $t/2$.

Proof. Let us consider the risk

$$R(\hat{\mu}(X); \mu) = E_\mu \|\hat{\mu}(X) - \mu\|^2$$

of an estimator of μ of the form

$$\hat{\mu}(X) = \phi(S)X, \tag{4.12}$$

where $\phi \colon \mathbb{R} \to \mathbb{R}$. Presuming the risk of $\hat{\mu}(X)$ to be everywhere finite or, equivalently, that

$$E_\mu\big(\phi^2(S)S\big) < \infty \quad \text{for all } \mu \in M, \tag{4.13}$$

we may write

$$R(\hat{\mu}(X); \mu) = E_\mu\big(\phi^2(S)S\big) - 2E_\mu\big(\phi(S)\langle X, \mu\rangle\big) + \|\mu\|^2. \tag{4.14}$$

By Proposition 3.8.7,

$$S = \|X\|^2 \sim \chi^2_{p; \|\mu\|} = \sum_{0 \leq k < \infty} p(k; \rho)\, \chi^2_{p+2k},$$

where

$$p(k; \rho) = e^{-\rho} \frac{\rho^k}{k!} \quad \text{with} \quad \rho = \frac{\|\mu\|^2}{2}.$$

The first term on the right side of (4.14) is thus

$$E_\mu\big(\phi^2(S)S\big) = \sum_{0 \leq k < \infty} p(k; \rho) \int_0^\infty \phi^2(s)s \,(\text{density of } \chi^2_{p+2k})(s)\, ds$$
$$= \sum_{0 \leq k < \infty} p(k; \rho)\, E\big(\phi^2(S_k)S_k\big), \tag{4.15}$$

where, for $k \geq 0$,

$$S_k \text{ denotes a random variable with a } \chi^2_{p+2k} \text{ distribution.} \tag{4.16}$$

To calculate the second term on the right of (4.14), let $e_1, \ldots, e_p$ be an orthonormal basis for M with e_1 pointing in the direction of μ, that is,

$$e_1 = \mu/u\,,$$

where

$$u = \|\mu\|\,.$$

For $x \in M$ write x_i for $\langle x, e_i \rangle$; similarly write X_i for $\langle X, e_i \rangle$. Since the X_i's are independent unit normal random variables with means $E(X_i) = \delta_{i1}u$,

$$E_\mu\big(\phi(S)\langle X, \mu \rangle\big) = u E_\mu\Big(\phi\big(\sum X_i^2\big)X_1\Big)$$

$$= u\,\frac{1}{(2\pi)^{p/2}} \int x_1 \phi\big(\sum x_i^2\big) \exp\Big[-\tfrac{1}{2}\sum x_i^2 + x_1 u - \tfrac{1}{2}u^2\Big]\,dx_1 \cdots dx_p$$

$$= u e^{-u^2/2}\,\frac{1}{(2\pi)^{p/2}} \int \frac{\partial}{\partial u}\Big(\phi\big(\sum x_i^2\big)\exp\Big[-\tfrac{1}{2}\sum x_i^2 + x_1 u\Big]\Big)\,dx_1 \cdots dx_p$$

$$= u e^{-u^2/2}\,\frac{1}{(2\pi)^{p/2}}\,\frac{\partial}{\partial u} \int \phi\big(\sum x_i^2\big)\exp\Big[-\tfrac{1}{2}\sum x_i^2 + x_1 u\Big]\,dx_1 \cdots dx_p$$

$$= u e^{-u^2/2}\,\frac{\partial}{\partial u}\Big(e^{u^2/2} E_\mu \phi(S)\Big)$$

$$= u e^{-u^2/2}\,\frac{\partial}{\partial u} \sum_{0 \le k < \infty} \frac{(u^2/2)^k}{k!}\,E\phi(S_k)$$

$$= \sum_{0 \le k < \infty} p(k; \rho)\,2k\,E\phi(S_k);$$

in order to justify the interchange of $\int$ and $\frac{\partial}{\partial u}$, we assume

$$E_\mu\big(|\phi|(S)\big) < \infty \quad \text{for all } \mu \in M. \tag{4.17}$$

To summarize, supposing the integrability conditions (4.13) and (4.17) to be in effect, $\hat{\mu}(X)$ has finite risk given by

$$R\big(\hat{\mu}(X); \mu\big) = \sum_{0 \le k < \infty} p(k; \rho)\big(ES_k \phi^2(S_k) - 4k E\phi(S_k) + 2k\big) \tag{4.18}$$

with $\rho = \|\mu\|^2/2$ and S_k satisfying (4.16).

Now let us look at the special case

$$\phi(s) = 1 - \frac{c}{s}\,.$$

For such a ϕ, conditions (4.13) and (4.17) hold if (and only if) S and $1/S$ have finite expectations. But

$$E_\mu(S) = \sum_{0 \le k < \infty} p(k; \rho)\,ES_k = \sum_{0 \le k < \infty} p(k; \rho)\,(p + 2k)$$

is trivially finite and by (4.6)

$$E_\mu\left(\frac{1}{S}\right) = \sum_{0 \le k < \infty} p(k; \rho) \, E\left(\frac{1}{S_k}\right)$$

$$= \sum_{0 \le k < \infty} p(k; \rho) \, \frac{1}{p - 2 + 2k} = \mathcal{E}_p(u^2)$$

is finite because of the assumption $p \ge 3$. By (4.18),

$$R\left(\left(1 - \frac{c}{S}\right)X; \mu\right) = \sum_{0 \le k < \infty} p(k; \rho) \, E\left[S_k \phi^2(S_k) - 4k\phi(S_k) + 2k\right]$$

$$= \sum_{0 \le k < \infty} p(k; \rho) \, E\left[\left(S_k - 2c + \frac{c^2}{S_k}\right) + \left(-4k + \frac{4kc}{S_k}\right) + 2k\right]$$

$$= \sum_{0 \le k < \infty} p(k; \rho) \left[p - 2c\left(1 - \frac{2k}{p - 2 + 2k}\right) + \frac{c^2}{p - 2 + 2k}\right]$$

$$= p - 2c(p - 2)\mathcal{E}_p(u^2) + c^2 \mathcal{E}_p(u^2). \tag{4.19}$$

This expression is minimized uniformly in $u = \|\mu\|$ by

$$c = p - 2,$$

the minimum being

$$p - (p - 2)^2 \mathcal{E}_p(u^2) = p - (p - 2)^2 E_\mu\left(\frac{1}{S}\right). \qquad \blacksquare$$

4.20 Exercise [2]. Suppose ϕ satisfies conditions (4.13) and (4.17). For $s \ge 0$ put $\phi^+(s) = \max(\phi(s), 0)$. Show that $R(\phi^+(S)X; \mu) \le R(\phi(S)X; \mu)$ for all μ, with strict inequality if the set $\{s : \phi(s) < 0\}$ has positive Lebesgue measure.   ◇

Remarks. (1) From Jensen's inequality applied to the convex function

$$f(y) = \frac{1}{p - 2 + 2y}, \quad y \ge 0,$$

we obtain

$$\mathcal{E}_p(t) = E\left(\frac{1}{p - 2 + 2K}\right) = Ef(K) \ge f(EK)$$

$$= f\left(\frac{t}{2}\right) = \frac{1}{p - 2 + t}, \tag{4.21}$$

whence

$$R(\hat{\mu}_{JS}; \mu) = p - (p - 2)^2 \mathcal{E}_p\left(\|\mu\|^2\right) \le p - \frac{(p - 2)^2}{(p - 2) + \|\mu\|^2} < p. \tag{4.22}$$

Equality holds in (4.21) when $t = 0$, so

$$R(\hat{\mu}_{JS}; 0) = 2. \qquad (4.23)$$

As (4.22) shows, the Gauss-Markov estimator $X = P_M Y$ of μ is inadmissible with respect to mean square error for $p = \dim(M) \geq 3$; indeed, $P_M Y$ is dominated by the *James-Stein estimator*

$$\hat{\mu}_{JS} = \left(1 - \frac{p-2}{S}\right) X. \qquad (4.24)$$

By (4.22), the domination is considerable for p large and μ near 0.

(2) Consider again the Bayesian framework where

$$\mu \sim N_M(0, \lambda I_M).$$

By (3.13) the Bayes rule $(1 - B(\lambda))X \equiv (1 - \frac{1}{1+\lambda})X$ has Bayes risk $p\frac{\lambda}{1+\lambda}$, which is an improvement of

$$p - p\frac{\lambda}{1 + \lambda} = \frac{p}{1 + \lambda}$$

over the Bayes risk p of the GME X. On the other hand, since $S \sim (1 + \lambda)\chi_p^2$ by (4.4), the Bayes risk of $\hat{\mu}_{JS} = (1 - \hat{B}(S))X \equiv (1 - \frac{p-2}{S})X$ is

$$E\left(p - (p-2)^2 E_\mu\left(\frac{1}{S}\right)\right) = p - (p-2)^2 E\left(\frac{1}{S}\right) = p - \frac{(p-2)^2}{(p-2)(1+\lambda)},$$

which is an improvement of

$$\frac{p-2}{1+\lambda}$$

over the Bayes risk of the GME X. Thus, from the perspective of improvement in Bayes risk, the James-Stein rule in fact compares very favorably with the Bayes rule it was designed to emulate; moreover, it has the advantage of not requiring foreknowledge of the diffuseness parameter λ.

(3) $\hat{\mu}_{JS}$ is minimax, since it dominates the minimax estimator $X = P_M Y$. $\hat{\mu}_{JS}$ is not, however, admissible, since by Exercise 4.20 it is itself dominated by the so-called *positive-part estimator*

$$\hat{\hat{\mu}}_{JS} \equiv \left(1 - \frac{p-2}{S}\right)^+ X = \max\left(0, 1 - \frac{p-2}{S}\right)X, \qquad (4.25)$$

which, unlike $\hat{\mu}_{JS}$, never reverses the sign of X in estimating μ. It is known that every admissible rule in the problem at hand must be a smooth function of the data, so $\hat{\hat{\mu}}_{JS}$ is also inadmissible; a dominating rule has yet to be found.

4.26 Figure. *For the case* $p = 10$, *the following diagram displays the risk functions of the GME* $\hat{\mu}_{GME} = X = P_M Y$, *the James-Stein estimator* $\hat{\mu}_{JS}$ *(4.24), the positive part estimator* $\hat{\hat{\mu}}_{JS}$ *(4.25), and the generalized Bayes estimator* $\hat{\mu}_a$ *of the next section for the cases* $a = 1/2$ *and* $a = 2$ *(see (5.13)). The risk functions depend on* μ *only through* $\|\mu\|$, *which is given on the horizontal axis. Evidently all four of the alternatives to* $\hat{\mu}_{GME}$ *have substantially smaller risks than* $\hat{\mu}_{GME}$ *for* μ's *near 0, but there is not much difference in the risks of the alternative estimators. The Bayes estimator* $\hat{\mu}_{a=1/2}$ *is somewhat preferable to* $\hat{\mu}_{JS}$ *and* $\hat{\hat{\mu}}_{JS}$ *in the region where those estimators provide a considerable reduction in risk.* $\hat{\mu}_{a=2}$ *dominates* $\hat{\hat{\mu}}_{JS}$.

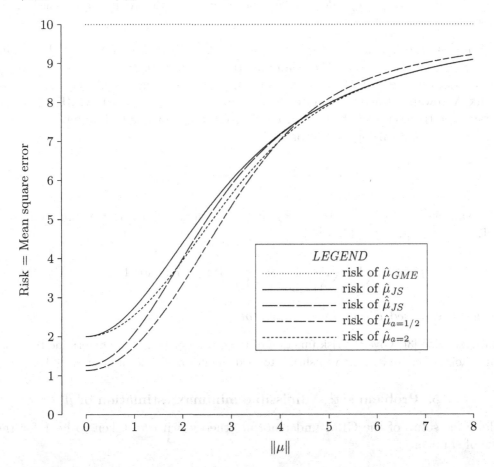

4.27 Exercise [3]. Suppose σ^2 is unknown. Let $\hat{\sigma}^2 = \|Q_M Y\|^2 / d(M^\perp)$ be the usual unbiased estimator of σ^2. Show that, with respect to normalized mean square error

$$R(\hat{\mu}; \mu, \sigma^2) = \frac{E_{\mu, \sigma^2} \|\hat{\mu} - \mu\|^2}{\sigma^2},$$

the best estimator of μ of the form

$$\left(1 - \frac{c\hat{\sigma}^2}{\|P_M Y\|^2}\right) P_M Y$$

is

$$\left(1 - \frac{q}{q+2} \frac{(p-2)\hat{\sigma}^2}{\|X\|^2}\right)X,$$

with risk

$$p - \frac{q}{q+2}(p-2)^2 \mathcal{E}_p\left(\frac{\|\mu\|^2}{\sigma^2}\right);$$

in these expressions $p = \dim(M) \geq 3$ and $q = \dim(M^\perp)$.

[Hint: In view of the independence of $\hat{\sigma}^2$ and $X = P_M Y$, this result follows easily from (4.19).] $\diamond$

The estimators considered so far have "shrunk" the GME X toward the origin, in accordance with the semi-Bayesian idea that μ is near 0. In some instances one's prior belief may be only that μ lies near some given subspace L of M. One way to shrink X towards L is to estimate the component of μ in L by its GME $P_L Y$ and to estimate the component of μ in $M - L$ using the previous techniques as applied to $P_{M-L}Y$; the resulting estimator is then

$$\hat{\mu} = P_L Y + \left(1 - \frac{c}{\|P_{M-L}Y\|^2}\right)P_{M-L}Y. \tag{4.28}$$

For example, when $V = M = \mathbb{R}^p$ and L is spanned by the equiangular line $(1, 1, \ldots, 1)^T$, equation (4.28) reads

$$\hat{\mu}_i = \bar{Y} + \left(1 - \frac{c}{\sum_{1 \leq j \leq p}(Y_j - \bar{Y})^2}\right)(Y_i - \bar{Y}) \quad \text{for } 1 \leq i \leq p; \tag{4.29}$$

this is the so-called *Efron-Morris estimator*.

4.30 Exercise [3]. Give the risk function of the estimator (4.28) and determine the optimum choice of c. How might you shrink toward the flat $m + L$ for a given $m \in M$? $\diamond$

5. Problem set: Admissible minimax estimation of μ

Adopt the setup of the GLM under normal theory, with σ^2 taken to be 1 for the sake of simplicity:

$$Y \sim N_V(\mu, I_V), \quad \mu \in M, \ \mu \text{ unknown.}$$

Suppose throughout that

$$p \equiv \dim(M) \geq 3.$$

The end result of this problem set is an admissible estimator $\hat{\mu} \equiv \hat{\mu}(Y)$ of μ dominating the GME $P_M Y$ with respect to the mean square error

$$R(\hat{\mu}; \mu) = E_\mu(\|\hat{\mu} - \mu\|^2) = \int_V \|\hat{\mu}(y) - \mu\|^2 \, \mathfrak{n}_\mu(y) \, dy,$$

where

$$\mathfrak{n}_\mu(y) = \frac{1}{(2\pi)^{\dim(V)/2}} \exp\left(-\tfrac{1}{2}\|y - \mu\|^2\right). \tag{5.1}$$

Since

$$X = P_M Y \sim N_M(\mu, I_M)$$

is a sufficient statistic, there is no real loss of generality in supposing, as we do, that $V = M$ and $Y = X$.

We will be considering estimators of μ of the form

$$\hat{\mu}(X) = \left(1 - \frac{(p-2)\tau(S)}{S}\right) X$$

$$= \tau(S)\left(\left(1 - \frac{p-2}{S}\right) X\right) + (1 - \tau(S)) X, \tag{5.2}$$

where

$$S = \|X\|^2$$

and $\tau\colon [0, \infty) \to \mathbb{R}$. By way of motivation, we note that such estimators are linear combinations (with coefficients depending on $\tau(S)$) of the James-Stein estimator $\hat{\mu}_{JS}$ and the Gauss-Markov estimator X; moreover, every estimator $\hat{\mu}$ of μ that is *spherically symmetric*, in the sense that $\hat{\mu}(\mathcal{O}X) = \mathcal{O}\hat{\mu}(X)$ for all orthogonal transformations $\mathcal{O}\colon V \to V$, must be of the form (5.2).

Our first objective is formula (5.8) below for the risk of such an estimator. Unlike formula (4.18), formula (5.8) gives a way of estimating the risk $R(\hat{\mu}(X); \mu)$ of $\hat{\mu}(X)$ directly from X. We begin with a preliminary technical result. Let $\mathcal{F}$ (respectively, $\mathcal{F}_+$) be the collection of functions $f\colon \mathbb{R} \to \mathbb{R}$ (respectively, $f\colon [0, \infty) \to \mathbb{R}$) that are continuous and have a piecewise continuous derivative.

A. Prove

5.3 Lemma (Stein). *Let $f \in \mathcal{F}$ and let Z be a normal random variable with mean θ and unit variance. If*

$$f'(Z) \quad \text{and} \quad (Z - \theta)f(Z) \quad \text{have finite expectations,}$$

then

$$E\big(f'(Z)\big) = E\big((Z - \theta)f(Z)\big). \tag{5.4}$$

[Hint: Integrate by parts.]   ○

Now we turn to

5.5 Theorem (Stein-Efron-Morris). *If $\tau \in \mathcal{F}_+$ and if*

$$E_\mu \frac{|\tau(S)|}{S} < \infty, \quad E_\mu \frac{\tau^2(S)}{S} < \infty, \quad \text{and} \quad E_\mu |\tau'(S)| < \infty, \tag{5.6}$$

then the risk $R(\hat{\mu}(X); \mu)$ of the estimator

$$\hat{\mu}(X) = \left(1 - \frac{(p-2)\tau(S)}{S}\right) X \equiv (1 - B(S)) X$$

is finite and estimated unbiasedly by

$$\hat{R}(S) \equiv p - (p-2)\left[(p-2)\left(\frac{\tau(S)(2 - \tau(S))}{S}\right) + 4\tau'(S)\right], \qquad (5.7)$$

that is,

$$R(\hat{\mu}(X); \mu) = E_\mu(\hat{R}(S)). \qquad (5.8)$$

Here is a sketch of the proof, setting aside questions of integrability. One has, writing simply E in place of E_μ,

$$\begin{aligned}
R(\hat{\mu}(X); \mu) &= E\big(\, \|(1 - B(S)) X - \mu\|^2 \,\big) \\
&= E\big(\, \|X - \mu\|^2 - 2\langle X - \mu, B(S)X \rangle + \|B(S)X\|^2 \,\big) \\
&= p - 2 \sum_{1 \le i \le p} E\big(B(S) X_i (X_i - \mu_i)\big) + E\big(B^2(S)S\big),
\end{aligned}$$

where, for some fixed orthonormal basis $e_1, \ldots, e_p$ for M,

$$X_i = \langle X, e_i \rangle \quad \text{and} \quad \mu_i = \langle \mu, e_i \rangle \quad \text{for } i = 1, \ldots, p.$$

The conditional expectation of $B(S)(X_i - \mu_i)X_i$, given $X_1 = x_1, \ldots, X_{i-1} = x_{i-1}$, $X_{i+1} = x_{i+1}, \ldots, X_p = x_p$, is

$$E\left[\left(B\big(\sum_{j \ne i} x_j^2 + X_i^2\big) X_i\right)(X_i - \mu_i)\right],$$

which by Stein's lemma is the same as

$$E\left[B\big(\sum_{j \ne i} x_j^2 + X_i^2\big) + 2B'\big(\sum_{j \ne i} x_j^2 + X_i^2\big) X_i^2\right].$$

Consequently

$$R(\hat{\mu}(X); \mu) = p - 2pE\big(B(S)\big) - 4E\big(B'(S)S\big) + E\big(B^2(S)S\big);$$

expressing $B(S)$ and $B'(S)$ in terms of S, $\tau(S)$, and $\tau'(S)$, one arrives at (5.8).

B. Complete the preceding argument, making it rigorous in the process.
[Hint: You will need the following facts about the integrability of random variables, say F and G, defined on a common probability space:

(a) If F and G are each integrable (that is, have finite expectations), then so is $F + G$, and
$$E(F + G) = E(F) + E(G).$$

(b) If F^2 and G^2 are each integrable, then so is FG.

(c) If F is integrable, then the conditional expectation, $E(F \mid G = g)$, of F given $G = g$ is defined for (almost) all g and is integrable in g with
$$E(F) = \int E(F \mid G = g) \, P(G \in dg).$$

(d) If $|F| \le |G|$ and G is integrable, then so is F.]   ○

C. Show that the integrability condition (5.6) in the Stein-Efron-Morris theorem is met if
$$\int_0^\infty |\tau'(s)| \, ds < \infty. \tag{5.9}$$

For $t \in \mathbb{R}$, use the Stein-Efron-Morris theorem to verify that the risk of
$$\hat{\mu}_{JS;t}(X) \equiv \left(1 - \frac{(p-2)t}{S}\right)X$$
is unbiasedly estimated by
$$\hat{R}_{JS;t}(S) \equiv p - (p-2)^2 \frac{t(2-t)}{S}$$
(compare (4.10)) and the risk of
$$\hat{\mu}_{JS;t}(X) \equiv \left(1 - \frac{(p-2)t}{S}\right)^+ X = \left[1 - \frac{(p-2)\min\left(t, S/(p-2)\right)}{S}\right]X$$
is unbiasedly estimated by
$$\hat{\hat{R}}_{JS;t}(S) = \begin{cases} \hat{R}_{JS;t}(S), & \text{if } S \ge (p-2)t, \\ S - p, & \text{if } S < (p-2)t. \end{cases} \qquad ○$$

D. Use the Stein-Efron-Morris theorem to deduce

5.10 Theorem (Baranchik). If $0 \le \tau \le 2$ and τ is nondecreasing, then $\hat{\mu}(X)$ defined by (5.2) is minimax.

[Hint: For $\tau \in \mathcal{F}_+$ this follows from (5.7)–(5.8). For a nonsmooth τ, first approximate $\tau(s)$ by
$$\tau_n(s) \equiv \int_{-\infty}^\infty \frac{1}{\sqrt{2\pi/n}} \exp\left(-\frac{(s-t)^2}{2/n}\right) \tau(t) \, dt$$
(put $\tau(t) = 0$ for $t \le 0$) and then pass to the limit as $n \to \infty$.]   ○

Now let Π be a prior distribution on μ. By (3.6), the Π-Bayes estimator of μ is $\hat{\mu}_{\Pi}(X)$, where

$$\hat{\mu}_{\Pi}(x) = \frac{\int_{\mu \in M} \mu \, \mathfrak{n}_{\mu}(x) \, \Pi(d\mu)}{\int_{\mu \in M} \mathfrak{n}_{\mu}(x) \, \Pi(d\mu)} \tag{5.11}$$

and where $\mathfrak{n}_{\mu}$ is defined by (5.1) with $V = M$. Up to now we have thought of priors as having unit probability mass. It is clear though that $\hat{\mu}_{\Pi}$ is well defined for any nonnegative measure Π of nonzero finite mass; in fact such a measure could be renormalized to unit mass without changing the value of $\hat{\mu}_{\Pi}$. There are technical advantages in considering $\hat{\mu}_{\Pi}$ even for nonnegative measures Π of infinite mass, subject to the proviso that the integrals appearing in (5.11) are finite for all x. When Π has finite mass, $\hat{\mu}_{\Pi}(X)$ is called a *proper* Bayes estimator. For Π of infinite mass, $\hat{\mu}_{\Pi}(X)$ is called a *formal*, or *improper*, Bayes estimator. Both cases are subsumed under the term *generalized* Bayes. In what follows we will be considering the generalized Bayes estimators for priors Π having the following structure. For a given $a \in \mathbb{R}$, one first selects β in the interval $(0,1)$ at random according to the density

$$\frac{d\beta}{\beta^a} \tag{5.12_1}$$

and then selects μ at random in M according to

$$N_M(0, \lambda I_M), \tag{5.12_2}$$

where

$$\lambda = \frac{1}{\beta} - 1. \tag{5.12_3}$$

Denote the resulting distribution of μ by Π_a. Note that the bigger is a, the more $d\beta/\beta^a$ favors smaller values of β and hence larger values of λ. Note also that when $a \geq 1$, Π_a has infinite mass.

E. Show that the generalized Bayes estimator versus the prior Π_a defined by (5.12) is well defined for each

$$a < 1 + \frac{p}{2}$$

and is given by the formula

$$\hat{\mu}_a(X) \equiv \hat{\mu}_{\Pi_a}(X) = \big(1 - B_a(S)\big) X, \tag{5.13}$$

where

$$B_a(s) = \frac{\int_0^1 \beta^{p/2-a+1} e^{-\beta s/2} \, d\beta}{\int_0^1 \beta^{p/2-a} e^{-\beta s/2} \, d\beta} = \frac{r}{\rho} \frac{P[G_{r+1,\rho} \leq 1]}{P[G_{r,\rho} \leq 1]} \tag{5.14_1}$$

$$= \frac{1}{s}\left(p + 2 - 2a - \frac{2 e^{-s/2}}{\int_0^1 \beta^{p/2-a} e^{-\beta s/2} \, d\beta}\right); \tag{5.14_2}$$

in (5.14_1) $G_{r,\rho}$ denotes a gamma random variable with shape parameter $r = p/2 - a + 1$ and rate parameter $\rho = s/2$.
[Hint: Use the results of Example 3.11.]   ○

F. The final result provides explicit admissible minimax estimators of μ for each p (≥ 3). Prove

5.15 Theorem (Strawderman-Berger). *Let* $a < 1 + p/2$ *and let* $\hat{\mu}_a(X)$ *be defined by* (5.13). *Then*

(i) $\hat{\mu}_a$ *is minimax if* $3 - p/2 < a$,

(ii) $\hat{\mu}_a$ *is admissible if* $a \leq 2$,

(iii) $\hat{\mu}_a$ *is proper Bayes if* $a < 1$.

[Hint: (i) follows from Baranchik's theorem. (iii) is trivial and covers the case $a < 1$ in (ii). For the rest of (ii), use the following result of Larry Brown (1971): If $p \geq 1$ and Π is spherically symmetric, in the sense that $\Pi(A) = \Pi(\mathcal{O}A)$ for all (measurable) sets A of M and all orthogonal transformations $\mathcal{O}: M \to M$, then $\hat{\mu}_\Pi(X)$ is admissible if

$$\sup_{x \in M} \|\hat{\mu}_\Pi(x) - x\| < \infty$$

and

$$\int_1^\infty \frac{dt}{t^{p-1}\phi(t)} = \infty,$$

where

$$\phi(t) = \int_{\mu \in M} \mathfrak{n}_\mu(x)\,\Pi(d\mu)$$

is the (generalized) marginal density of X at any point $x \in M$ with $\|x\| = t$.]   ○

CHAPTER 6

NORMAL THEORY: TESTING

Suppose V, $\langle \cdot, \cdot \rangle$ is an inner product space. *Throughout this chapter*, we assume that

$$Y \sim N_V(\mu, \sigma^2 I_V) \qquad (0.1)$$

with μ (unknown) lying in a given proper subspace M of V and with σ^2 (unknown) > 0. We are interested in the problem of testing

$$H: \mu \in M_0 \quad \text{versus} \quad A: \mu \notin M_0, \qquad (0.2)$$

where M_0 is a given subspace of M.

0.3 Example *(The triangle problem).* As in Example 4.1.5, let $V = \mathbb{R}^3$ and $E(Y_i) = \beta_i$ for $i = 1$, 2, and 3 with $\beta_1 + \beta_2 + \beta_3 = 0$, so

$$M = \left\{ \sum_{1 \leq j \leq 3} \beta_j e^{(j)} : \beta_1 + \beta_2 + \beta_3 = 0 \right\}$$

with $(e^{(j)})_i = \delta_{ij}$. Thinking of the β_j's as the angles, less 60°, of a triangle, the hypothesis that "the angles are equal" is expressed as $\beta_1 = \beta_2 = \beta_3$ or as $\beta_1 = \beta_2 = \beta_3 = 0$ in view of the constraint. The "equilateral hypothesis" therefore corresponds to $\mu \in M_0$, where

$$M_0 = 0. \qquad \bullet$$

0.4 Example *(Simple linear regression).* Here $V = \mathbb{R}^n$ and $EY_i = \alpha + \beta x_i$ for $1 \leq i \leq n$, so

$$M = [e, x] \quad \text{with} \quad (x)_i = x_i.$$

The hypothesis that the slope β is 0 corresponds to $\mu \in M_0$ with

$$M_0 = [e]. \qquad \bullet$$

0.5 Example *(One-way analysis of variance).* As in Example 4.2.2(C), suppose $V = \left\{ (x_{ij})_{j=1,\ldots,n_i; i=1,\ldots,p} \in \mathbb{R}^n \right\}$ with $n = \sum_i n_i$ and $E(Y_{ij}) = \beta_i$ for $j = 1, \ldots, n_i$,

$i = 1, \ldots, p$, so

$$M = \left\{ \sum_{1 \leq i \leq p} \boldsymbol{v}^{(i)} \beta_i : \beta_1, \ldots, \beta_p \in \mathbb{R} \right\}$$

with $\left(\boldsymbol{v}^{(i)} \right)_{i'j'} = \delta_{ii'}$. The hypothesis that the group means β_i are equal corresponds to $\mu \in M_0$ with

$$M_0 = \left[\sum_{1 \leq i \leq p} \boldsymbol{v}^{(i)} \right] = [e]. \qquad \bullet$$

0.6 Exercise [1]. Which of the following hypotheses fall directly within the context of (0.2) or can be made to do so by a simple modification?

  (a) In the triangle problem, the hypothesis that angles 1 and 2 are equal.
  (b) In the triangle problem, the hypothesis that the triangle is isosceles.
  (c) In simple linear regression, the hypothesis that the slope is 1.
  (d) In simple linear regression, the hypothesis that the slope is positive.
  (e) In one-way ANOVA, the hypothesis that $\beta_1 < \beta_2 < \cdots < \beta_p$.
  (f) In one-way ANOVA, the hypothesis that $(\beta_1 + \beta_2)/2 = \beta_3$.
  (g) In one-way ANOVA, the hypothesis that $\beta_1 = \beta_2 = \beta_3$ and $\beta_4 = \beta_5 = \cdots = \beta_p$. $\diamond$

In the next two sections we show that both the likelihood ratio principle and several appealing heuristic arguments point to the classical F-test as the "right" test of (0.2). In Section 6.3, the power of the F-test is shown to be a nondecreasing function of the distance from $\mu \in M$ to M_0 in σ units; this fact is used in Section 4 in establishing the optimality of the F-test within a certain class of invariant tests. Section 5 shows how the problem of simultaneous inference on linear functionals of μ relates to the F-test. The chapter closes with a problem set that examines the optimality of the F-test with respect to a criterion involving average power.

1. The likelihood ratio test

We will derive the likelihood ratio test of (0.2). By (3.7.6), Y has density

$$f_{\mu,\sigma^2}(y) = \frac{1}{(2\pi\sigma^2)^{n/2}} e^{-\frac{1}{2}\|y-\mu\|^2/\sigma^2} \quad (n = \dim(V))$$

with respect to Lebesgue measure on V. The likelihood ratio test employs the criterion

$$\lambda(Y) = \frac{\sup_{\mu \in M_0, \sigma^2 > 0} f_{\mu,\sigma^2}(Y)}{\sup_{\mu \in M, \sigma^2 > 0} f_{\mu,\sigma^2}(Y)}$$

together with the rejection region

$$\lambda(Y) \leq c,$$

where c is chosen to give the test the desired size. For notational convenience, set $M_1 = M$. For $i = 0$ or 1, $\mu \in M_i$, and $\sigma^2 > 0$ one has

$$f_{\mu,\sigma^2}(Y) = \frac{1}{(2\pi\sigma^2)^{n/2}} e^{-\frac{1}{2}\|P_{M_i}Y-\mu\|^2/\sigma^2} e^{-\frac{1}{2}\|Q_{M_i}Y\|^2/\sigma^2}.$$

This expression is a maximum at the maximum likelihood estimators $\hat{\mu}_{\text{MLE}} = P_{M_i}Y$ and $\hat{\sigma}^2_{\text{MLE}} = \|Q_{M_i}Y\|^2/n$, the maximum being

$$f_{\hat{\mu}_{\text{MLE}}, \hat{\sigma}^2_{\text{MLE}}}(Y) = \frac{1}{(2\pi)^{n/2}} \left(\frac{n}{\|Q_{M_i}Y\|^2} \right)^{n/2} e^{-n/2}.$$

Thus

$$\lambda(Y) = \left(\frac{\|Q_MY\|^2}{\|Q_{M_0}Y\|^2} \right)^{n/2}$$

and the likelihood ratio test rejects for small values of the ratio of the distance of Y from M to the distance of Y from the smaller subspace M_0.

2. The F-test

It is convenient to put the likelihood ratio test in an equivalent form in which the distributions of the test statistic are more readily dealt with. We do this now:

$$\text{the LHR test rejects} \iff \frac{\|Q_MY\|^2}{\|Q_{M_0}Y\|^2} \text{ is small}$$

$$\iff \frac{\|Q_{M_0}Y\|^2}{\|Q_MY\|^2} \text{ is large}$$

$$\iff \frac{\|Q_{M_0}Y\|^2 - \|Q_MY\|^2}{\|Q_MY\|^2} \text{ is large}$$

$$\iff F \equiv \frac{\|P_{M-M_0}Y\|^2/d(M - M_0)}{\|Q_MY\|^2/d(M^\perp)} \text{ is large;} \tag{2.1}$$

at the last step we used the Pythagorean identity

$$\|Q_{M_0}Y\|^2 = \|Q_MY\|^2 + \|P_{M-M_0}Y\|^2.$$

By Theorem 3.8.2, for any $\mu \in M$ and $\sigma^2 > 0$,

$$\|P_{M-M_0}Y\|^2 \sim \sigma^2 \chi^2_{d(M-M_0), \|P_{M-M_0}\mu\|/\sigma}$$

independently of

$$\|Q_MY\|^2 \sim \sigma^2 \chi^2_{d(M^\perp)}.$$

It follows that when μ and σ^2 obtain, the distribution of F is the (noncentral) $\mathcal{F}_{d(M-M_0), d(M^\perp); \gamma}$ distribution, the noncentrality parameter γ being given by

$$\gamma = \frac{\|P_{M-M_0}\mu\|}{\sigma} = \frac{\|\mu - P_{M_0}\mu\|}{\sigma} = \frac{\|Q_{M_0}\mu\|}{\sigma}; \tag{2.2}$$

note that γ is the distance from μ to M_0, in units of σ. Under the hypothesis H that μ lies in M_0, one has $\gamma = 0$, and so to achieve size α one rejects for

$$F \geq c, \tag{2.3}$$

where c is the upper α fractional point of the central $\mathcal{F}_{d(M-M_0),d(M^\perp)}$ distribution. This is the so-called F-test (of size α). To summarize:

2.4 Proposition. For the testing problem (0.1)–(0.2), the likelihood ratio test (of size α) coincides with the F-test (of size α).

In view of the orthogonality of M_0, $M - M_0$, and $M^\perp$, the quantities figuring in the F-statistic (2.1) may be expressed in various ways:

$$d(M^\perp) = d(V) - d(M),$$
$$d(M - M_0) = d(M) - d(M_0),$$
$$\|P_{M-M_0}Y\|^2 = \|P_M Y\|^2 - \|P_{M_0}Y\|^2 = \|Q_{M_0}Y\|^2 - \|Q_M Y\|^2,$$
$$\|Q_M Y\|^2 = \|Y\|^2 - \|P_M Y\|^2,$$
$$\|Q_{M_0}Y\|^2 = \|Y\|^2 - \|P_{M_0}Y\|^2.$$

The following schematic diagram may be helpful in thinking about these identities; M_0 lies along the horizontal axis, $M^\perp$ along the vertical axis, and $M - M_0$ along an axis receding into the page.

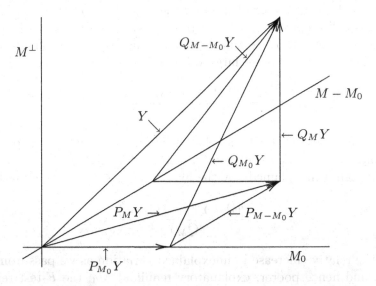

$\|P_{M-M_0}Y\|^2$ is often called the *sum of squares for testing H*, denoted SS_H, and $\|Q_M Y\|^2$ is often called the *sum of squares for error*, denoted SS_e. The terminology comes from the case of $(V, \langle\cdot,\cdot\rangle) = (\mathbb{R}^n,$ dot product$)$, where the squared length of a vector is the sum of squares of its components.

The F-test has the following simple interpretations:

A. For any $\mu \in M$ and $\sigma^2 > 0$, formula (3.5.5) for the expected squared length of a weakly spherical random vector implies that

$$\hat{\sigma}_H^2 \equiv \|P_{M-M_0}Y\|^2/d(M - M_0)$$

has expectation

$$E(\hat{\sigma}_H^2) = \sigma^2 + \frac{\|Q_{M_0}\mu\|^2}{d(M - M_0)}$$

while

$$\hat{\sigma}^2 \equiv \|Q_M Y\|^2 / d(M^\perp)$$

has expectation

$$E(\hat{\sigma}^2) = \sigma^2.$$

When $H: \mu \in M_0$ is true, both $\hat{\sigma}_H^2$ and $\hat{\sigma}^2$ unbiasedly estimate σ^2 and one expects their ratio

$$F = \frac{\hat{\sigma}_H^2}{\hat{\sigma}^2}$$

to be around one. When H is false, $\hat{\sigma}^2$ still unbiasedly estimates σ^2 while $\hat{\sigma}_H^2$ tends to overestimate σ^2, so one expects F to be larger than one. Hence it is reasonable to reject H for large values of F.

B. $P_M Y$ is the best estimator of μ under the general assumptions ($\mu \in M$), while $P_{M_0} Y$ is the best estimator of μ under the hypothesis H ($\mu \in M_0$). The F-test rejects if the discrepancy $\|P_M Y - P_{M_0} Y\|$ between these two estimators of μ is too large compared to the estimate $\hat{\sigma}$ of the underlying variability in the data; that is, the F-test rejects for large values of

$$\frac{\|P_M Y - P_{M_0} Y\|}{\hat{\sigma}}.$$

C. Suppose we regard $\|Q_M Y\|^2$ as the variability in Y left "unexplained" after taking into account the assumption $\mu \in M$, and $\|Q_{M_0} Y\|^2$ as the corresponding variability left "unexplained" by the hypothesis $\mu \in M_0$. The quantity

$$\frac{\|Q_{M_0} Y\|^2 - \|Q_M Y\|^2}{\|Q_M Y\|^2}$$

is then the relative increase in unexplained variation as we pass from M to the smaller, and hence poorer, explanatory manifold M_0; the F-test rejects if the increase is too large.

2.5 Exercise [2]. When $H: \mu \in M_0$ is true,

$$\hat{\sigma}_0^2 \equiv \frac{\|Q_{M_0} Y\|^2}{d(M_0^\perp)}$$

unbiasedly estimates σ^2, and more accurately so (there being more degrees of freedom) than the usual

$$\hat{\sigma}^2 = \frac{\|Q_M Y\|^2}{d(M^\perp)}.$$

It might be thought that a better test of $H\colon \mu \in M_0$ would be obtained by rejecting for large values of the statistic

$$\frac{\|P_{M-M_0}Y\|^2/d(M - M_0)}{\hat{\sigma}_0^2},$$

the null distribution of which does not depend on $\mu \in M_0$ or on σ^2. How does this test relate to the F-test? ◇

The *power* of the F-test is

$$\beta(\mu, \sigma^2) = P_{\mu,\sigma^2}[\text{reject } H] = \mathcal{F}_{d(M-M_0),d(M^\perp);\gamma}\big([c, \infty)\big), \qquad (2.6)$$

where the noncentrality parameter γ is given by (2.2) and c is the upper α point of the central $\mathcal{F}_{d(M-M_0),d(M^\perp)}$ distribution. Notice that

$$\frac{\sigma^2 \gamma^2}{d(M - M_0)} = \frac{\|P_{M-M_0}\mu\|^2}{d(M - M_0)} \qquad (2.7)$$

is just the numerator of the F-statistic evaluated for $Y = \mu$; this observation is helpful in obtaining γ. In the next section we will prove that the power of the F-test is an increasing function of γ.

2.8 Example *(The triangle problem).* Here $V = \mathbb{R}^3$ and $E(Y_i) = \beta_i$ for $i = 1$, 2, and 3 with $\beta_1 + \beta_2 + \beta_3 = 0$, so

$$M = [e]^\perp$$

and

$$M_0 = 0$$

for the equilateral hypothesis. One has

$$M^\perp = [e], \quad Q_M Y = \bar{Y}e, \quad d(M^\perp) = 1,$$
$$M - M_0 = M, \quad d(M - M_0) = 2,$$
$$P_{M-M_0}Y = P_M Y = Y - Q_M Y = Y - \bar{Y}e.$$

The F-statistic for testing $H\colon \mu \in M_0$ versus $A\colon \mu \notin M_0$ is thus

$$\frac{\|P_{M-M_0}Y\|^2/d(M - M_0)}{\|Q_M Y\|^2/d(M^\perp)} = \frac{\sum_{1 \le i \le 3}(Y_i - \bar{Y})^2/2}{3\bar{Y}^2/1},$$

distributed as $\mathcal{F}_{2,1;\gamma}$ with

$$\gamma = \frac{\sqrt{\sum_{1 \le i \le 3}(\beta_i - \bar{\beta})^2}}{\sigma} = \frac{\sqrt{\sum_{1 \le i \le 3}\beta_i^2}}{\sigma} = \frac{\|\mu\|}{\sigma}$$

$(\bar{\beta} \equiv (\beta_1 + \beta_2 + \beta_3)/3 = 0$ by the constraint). ●

2.9 Example *(Simple linear regression).* Here $V = \mathbb{R}^n$ and $E(Y_i) = \alpha + \beta x_i$ for $1 \leq i \leq n$, so

$$M = [e, x] = [e, v],$$

where $(x)_i = x_i$ and $v = Q_{[e]}x$. For the hypothesis of 0 slope

$$M_0 = [e].$$

One has

$$Q_M Y = Y - P_M Y = Y - P_{[e]}Y - P_{[v]}Y = Y - \bar{Y}e - \hat{\beta}v,$$
$$d(M^\perp) = n - 2,$$

with

$$\hat{\beta} = \mathrm{GME}(\beta) = \langle cv(\beta), Y \rangle = \langle \frac{v}{\|v\|^2}, Y \rangle = \frac{\sum_i Y_i(x_i - \bar{x})}{\sum_i (x_i - \bar{x})^2}$$

and

$$M - M_0 = [v], \quad P_{M-M_0}Y = \hat{\beta}v, \quad d(M - M_0) = 1.$$

The F-statistic is

$$\frac{\|P_{M-M_0}Y\|^2/d(M - M_0)}{\hat{\sigma}^2} = \frac{\hat{\beta}^2 \sum_i (x_i - \bar{x})^2/1}{\sum_i \left(Y_i - \bar{Y} - \hat{\beta}(x_i - \bar{x})\right)^2/(n - 2)},$$

distributed as $\mathcal{F}_{1,n-2;\gamma}$ with

$$\gamma = \frac{\sqrt{\beta^2 \sum_{1 \leq i \leq n}(x_i - \bar{x})^2}}{\sigma}.$$

$\bullet$

2.10 Exercise [2]. In the spirit of the two preceding examples, develop the F-test for the hypothesis posed in Example 0.5. $\diamond$

2.11 Exercise [3]. In the context of the GLM, suppose ψ is a given linear functional on M, $\hat{\psi}$ its GME, and $\hat{\sigma}_{\hat{\psi}} = \hat{\sigma}\|\psi\|$ its estimated standard error (see (4.4.2)). Show that the F-statistic for testing

$$H: \psi(\mu) = 0 \quad \text{versus} \quad A: \psi(\mu) \neq 0$$

is

$$\left(\frac{\hat{\psi}(Y)}{\hat{\sigma}_{\hat{\psi}}}\right)^2, \tag{2.12}$$

distributed as $\mathcal{F}_{1,d(M^\perp);\gamma}$ with

$$\gamma = \frac{|\psi(\mu)|}{\sigma_{\hat{\psi}}}. \tag{2.13}$$

Review Example 2.9 from this perspective. $\diamond$

3. Monotonicity of the power of the F-test

Here we will prove:

3.1 Proposition. *The power of the F-test is an increasing function of the noncentrality parameter*

$$\gamma = \frac{\|P_{M-M_0}\mu\|}{\sigma} .$$

In fact we will prove even more than this, namely, that for fixed m and n, the noncentral $\mathcal{F}$ family $\{\, \mathcal{F}_{m,n;\gamma} : \gamma \geq 0 \,\}$ has the so-called monotone likelihood ratio property with respect to γ; the desired monotonicity of the power of the F-test will follow as a simple corollary.

Let $(P_\theta)_{\theta \in \Theta}$ be a family of probability distributions on $\mathbb{R}$ indexed by a subset Θ of $\mathbb{R}$. This family is said to have a *monotone (increasing) likelihood ratio with respect to θ*, in brief, *MLR*, if the P_θ's have densities f_θ with respect to some measure ν on $\mathbb{R}$ such that, for all $\theta_1 < \theta_2$,

$$\lambda_{\theta_1,\theta_2}(x) \equiv \frac{f_{\theta_2}(x)}{f_{\theta_1}(x)} \tag{3.2}$$

is nondecreasing in x over the region in which it is not indeterminate $\left(= \frac{0}{0}\right)$. To put it another way, $(P_\theta)_{\theta \in \Theta}$ has an MLR if and only if, for all $\theta_1 < \theta_2$ and $x_1 < x_2$, one has

$$\frac{f_{\theta_2}(x_1)}{f_{\theta_1}(x_1)} \leq \frac{f_{\theta_2}(x_2)}{f_{\theta_1}(x_2)} \tag{3.3$_1$}$$

(with $\leq$ holding by convention if either side is $\frac{0}{0}$) or, equivalently,

$$\det \begin{pmatrix} f_{\theta_1}(x_1) & f_{\theta_1}(x_2) \\ f_{\theta_2}(x_1) & f_{\theta_2}(x_2) \end{pmatrix} \geq 0. \tag{3.3$_2$}$$

Since this condition is symmetric in x and θ, one says that $f_\theta(x)$ has *an MLR in x and θ*.

Examples of families $(P_\theta)_{\theta \in \Theta}$ having an MLR are the one-parameter exponential families with densities of the form

$$f_\theta(x) = C(\theta)\, h(x)\, e^{T(x)Q(\theta)}, \tag{3.4}$$

where $Q(\theta)$ is nondecreasing in θ and $T(x)$ is nondecreasing in x; indeed, for such a family,

$$\frac{f_{\theta_2}(x)}{f_{\theta_1}(x)} = \frac{C(\theta_2)}{C(\theta_1)}\, e^{T(x)[Q(\theta_2)-Q(\theta_1)]}$$

is nondecreasing in x for $\theta_1 < \theta_2$. Here are a couple of specific cases:

$1°$ The *Poisson family* $\{\, \mathcal{P}(\theta) : \theta \geq 0 \,\}$ with $\mathcal{P}(\theta)$ having density

$$p_\theta(x) = e^{-\theta} \frac{1}{x!} \theta^x = C(\theta)\, h(x)\, e^{x \log(\theta)}$$

with respect to counting measure on the nonnegative integers.

$2°$ The *unnormalized* $\mathcal{F}$ *family* $\{\, \mathcal{F}^*_{m,n} : m \geq 1 \,\}$ for fixed $n \geq 1$. $\mathcal{F}^*_{m,n}$ is the distribution of U/V for $U \sim \chi^2_m$, $V \sim \chi^2_n$, and U and V independent; it has density

$$f^*_{m,n}(x) = \frac{1}{\mathrm{Beta}(m/2, n/2)} \frac{x^{m/2-1}}{(1+x)^{(m+n)/2}}$$
$$= C(m)\, h(x)\, e^{\log[x/(1+x)]\, m/2}$$

with respect to Lebesgue measure on $[0, \infty)$.

Now consider the noncentral unnormalized $\mathcal{F}^*_{m,n;\gamma}$ distribution, defined to be the distribution of U/V for $U \sim \chi^2_{m;\gamma}$, $V \sim \chi^2_n$, and U and V independent.

3.5 Proposition. *For fixed m and n, the family $\{\, \mathcal{F}^*_{m,n;\gamma} : \gamma \geq 0 \}$ has an MLR in γ.*

Proof. By Exercise 3.8.13,

$$\mathcal{F}^*_{m,n;\gamma} = \sum_{0 \leq k < \infty} p_{\theta(\gamma)}(k)\, \mathcal{F}^*_{m+2k,n} \quad \text{with} \quad \theta(\gamma) = \frac{\gamma^2}{2},$$

whence $\mathcal{F}^*_{m,n;\gamma}$ has density

$$f^*_{m,n;\gamma}(x) = \sum_{0 \leq k < \infty} p_{\theta(\gamma)}(k)\, f^*_{m+2k,n}(x).$$

By $1°$ above, $p_{\theta(\gamma)}(k)$ has an MLR in γ and k while by $2°$ above, $f^*_{m+2k,n}(x)$ has an MLR in k and x. The proposition below guarantees that $f^*_{m,n;\gamma}(x)$ has an MLR in γ and x and hence that the family $\{\, \mathcal{F}^*_{m,n;\gamma} : \gamma \geq 0 \}$ has an MLR in γ. ∎

3.6 Exercise [2]. The noncentral normalized $\mathcal{F}_{m,n;\gamma}$ distribution is the distribution of $(U/m)/(V/n)$ for $U \sim \chi^2_{m;\gamma}$, $V \sim \chi^2_n$, and U and V independent. Show that for fixed m and n, the family $\{\, \mathcal{F}_{m,n;\gamma} : \gamma \geq 0 \,\}$ has an MLR in γ. ◇

3.7 Proposition. *Suppose $0 \leq p(x,t)$ has an MLR in x and t and $0 \leq q(t,y)$ has an MLR in t and y. Let τ be a nonnegative measure such that*

$$r(x,y) = \int p(x,t) q(t,y)\, \tau(dt) \tag{3.8}$$

is finite for each x and y. Then $r(x,y)$ has an MLR in x and y.

Proof. We have to show that when $x_1 < x_2$ and $y_1 < y_2$,

$$\frac{r(x_2, y_1)}{r(x_1, y_1)} \leq \frac{r(x_2, y_2)}{r(x_1, y_2)} \tag{3.9}$$

in the case that neither side above is of the form $\frac{0}{0}$. We will do this assuming

$$r(x_1, y_1) > 0,\ r(x_1, y_2) > 0,\ \text{and}$$
$$p(x_2, t) = 0 \text{ for each } t \text{ such that } p(x_1, t) = 0; \tag{3.10}$$

the general case is left to the reader as an exercise. For $y = y_1$ or y_2, write

$$\frac{r(x_2, y)}{r(x_1, y)} = \frac{\int \left(\frac{p(x_2,t)}{p(x_1,t)}\right) p(x_1,t) q(t,y)\, \tau(dt)}{\int p(x_1,t) q(t,y)\, \tau(dt)} = \int \phi(t) s_y(t)\, \tau(dt),$$

where

$$\phi(t) = \frac{p(x_2, t)}{p(x_1, t)}$$

and

$$s_y(t) = \frac{p(x_1, t) q(t, y)}{r(x_1, y)}\ .$$

Note that

$$0 \leq s_y(t) \quad \text{and} \quad \int s_y(t)\, \tau(dt) = 1,$$

so that the measure S_y having density s_y with respect to τ is a probability measure; moreover, because $q(t, y)$ has an MLR in t and $y = y_1, y_2$, the family $\{\,S_y : y = y_1, y_2\,\}$ has an MLR with respect to $y = y_1, y_2$. Because $p(x, t)$ has an MLR in $x = x_1, x_2$ and t, $\phi(t)$ is nondecreasing in t (off of the region $\{\,t : p(x_1, t) = 0 = p(x_2, t)\,\}$, which has probability 0 under S_{y_1} and S_{y_2}). That (3.9) holds now follows from Proposition 3.12 below.   ∎

3.11 Exercise [3]. Deduce (3.9) without assuming (3.10).
[Hint: Replace $p(x_1, t)$ by $p(x_1, t) + p(x_2, t)$, and hence $r(x_1, y_i)$ by $r(x_1, y_i) + r(x_2, y_i)$ for $i = 1, 2$.]   ◇

3.12 Proposition. *Let $(P_\theta)_{\theta \in \Theta}$ be a family of probabilities having an MLR. Let $\phi \colon \mathbb{R} \to \mathbb{R}$ be a nondecreasing function such that*

$$E_\theta \phi \equiv \int \phi(x)\, P_\theta(dx) \tag{3.13}$$

is finite for each θ. Then $E_\theta \phi$ is a nondecreasing function of θ.

Remarks. (1) If $(P_\theta)_{\theta \in \Theta}$ has an MLR, then for any $x_0 \in \mathbb{R}$

$$P_\theta([x_0, \infty)) = \int I_{[x_0, \infty)}(x)\, P_\theta(dx)$$

is nondecreasing in θ; this says that the mass of P_θ "moves to the right" as θ increases. This property is called *stochastic monotonicity*.

(2) The power of the F-test is

$$\beta(\gamma) = \mathcal{F}_{d(M-M_0),d(M^\perp);\gamma}\big([c,\infty)\big),$$

where c is chosen so that

$$\beta(0) = \alpha,$$

the prescribed level of significance. Since $\{\mathcal{F}_{m,n;\gamma} : \gamma \geq 0\}$ has an MLR in γ, $\beta(\gamma)$ is nondecreasing in γ, as asserted. In particular, $\beta(\gamma) \geq \alpha$ for all $\gamma > 0$, that is, the F-test is "unbiased."

Proof of Proposition 3.12. Let $\theta_1 < \theta_2$ and let f_{θ_1} and f_{θ_2} be the densities of P_{θ_1} and P_{θ_2} with respect to the base measure ν. Then

$$E_{\theta_2}\phi - E_{\theta_1}\phi = \int \phi(x)\big(f_{\theta_2}(x) - f_{\theta_1}(x)\big)\,\nu(dx) \equiv \int \phi(f_{\theta_2} - f_{\theta_1})\,d\nu$$

$$= \int_L \phi(f_{\theta_2} - f_{\theta_1})\,d\nu + \int_G \phi(f_{\theta_2} - f_{\theta_1})\,d\nu,$$

where

$$L = \{\,x : f_{\theta_2}(x) < f_{\theta_1}(x)\,\} \quad \text{and} \quad G = \{\,x : f_{\theta_2}(x) > f_{\theta_1}(x)\,\}.$$

Since $\lambda_{\theta_1,\theta_2}(x) = f_{\theta_2}(x)/f_{\theta_1}(x)$ is nondecreasing in x, L lies to the left of G. Because $\phi(x)$ is nondecreasing in x,

$$\sup_{x\in L} \phi(x) \equiv \ell \leq g \equiv \inf_{x\in G} \phi(x).$$

Thus

$$E_{\theta_2}\phi - E_{\theta_1}\phi \geq \ell \int_L (f_{\theta_2} - f_{\theta_1})d\nu + g \int_G (f_{\theta_2} - f_{\theta_1})d\nu$$

$$= (g - \ell) \int_G (f_{\theta_2} - f_{\theta_1})d\nu \geq 0,$$

with the equality on the preceding line holding because

$$\int_G (f_{\theta_2} - f_{\theta_1})d\nu + \int_L (f_{\theta_2} - f_{\theta_1})d\nu = \int (f_{\theta_2} - f_{\theta_1})d\nu$$

$$= \int f_{\theta_2}\,d\nu - \int f_{\theta_1}\,d\nu = 1 - 1 = 0. \qquad \blacksquare$$

3.14 Exercise [2]. Show that for fixed m the family $\{\chi^2_{m;\gamma} : \gamma \geq 0\}$ of noncentral χ^2 distributions has an MLR in γ. $\qquad\qquad\diamond$

3.15 Exercise [3]. The *unnormalized noncentral* $t^*_{n;\theta}$ distribution is the distribution of $U/\sqrt{V}$, where $U \sim N(\theta,1)$, $V \sim \chi^2_n$, and U and V are independent. Show that for each fixed n, the family $\{t^*_{n;\theta} : -\infty < \theta < \infty\}$ has an MLR in θ.
[Hint: The density of $t^*_{n;\theta}$ is

$$f^*_{n;\theta}(t) = \frac{1}{\sqrt{2\pi}\Gamma(n/2)2^{n/2}} \int_0^\infty e^{-\frac{1}{2}(t\sqrt{v}-\theta)^2} v^{\frac{n-1}{2}} e^{-\frac{v}{2}}\,dv. \tag{3.16}$$

When $t > 0$, the change of variables $w = t\sqrt{v}$ and Proposition 3.7 show that $f_{n;\theta}^*(t)$ has an MLR in θ and t. For $t < 0$ use $f_{n;\theta}^*(t) = f_{n;-\theta}^*(-t)$.]   ◇

3.17 Exercise [4]. Deduce Proposition 3.12 from Proposition 3.7.
[Hint: Think about $\int_{-\xi}^{\xi} e^{t\phi(x)} f_\theta(x)\, \nu(dx)$ for ξ near ∞ and t near 0.]   ◇

4. An optimal property of the F-test

We will show that the F-test is uniformly most powerful (UMP) in a certain class of invariant tests. We begin by describing the notion of invariance in the context of a general hypothesis test.

Let X be an $\mathfrak{X}$-valued random variable whose distribution belongs to some given class $(P_\theta)_{\theta \in \Theta}$ of distinct probabilities on the given observation space $\mathfrak{X}$. Consider testing

$$H: \theta \in \Theta_H \quad \text{versus} \quad A: \theta \in \Theta_A, \tag{4.1}$$

with Θ being the disjoint union of Θ_H and Θ_A. Let $g: \mathfrak{X} \to \mathfrak{X}$ be a *transformation* of $\mathfrak{X}$, that is, a one-to-one (bimeasurable) mapping of $\mathfrak{X}$ onto itself. g is said to *leave the testing problem invariant*, written $g \in \mathcal{I}$, if the possible distributions of $g(X)$ are precisely the possible distributions of X and if $g(X)$ is distributed according to a hypothesis distribution if and only if X is. Clearly the identity transformation, which maps each $x \in \mathfrak{X}$ to itself, leaves the problem invariant. Since the composition hg of two transformations g and h of $\mathfrak{X}$ acts on X by the rule

$$(hg)(X) = h\big(g(X)\big), \tag{4.2}$$

one has $hg \in \mathcal{I}$ whenever $g \in \mathcal{I}$ and $h \in \mathcal{I}$, and $g^{-1} \in \mathcal{I}$ whenever $g \in \mathcal{I}$. (g^{-1} denotes the inverse of g.) Consequently, if $\mathcal{C}$ is a class of transformations each of which leaves the problem invariant, then each element of the subgroup $\mathcal{G}$ (of all transformations of $\mathfrak{X}$) generated by $\mathcal{C}$ is in $\mathcal{I}$.

4.3 Example. Consider the setup of the GLM (but identify Y there with X here): $\mathfrak{X} = V$ and

$$X \sim N_V(\mu, \sigma^2 I_V) = P_{\mu,\sigma^2}$$
$$\Theta = \big\{ (\mu, \sigma^2) : \mu \in M, \sigma^2 > 0 \big\} = M \times (0, \infty)$$
$$\Theta_H = M_0 \times (0, \infty), \quad \Theta_A = \{\theta \in \Theta : \theta \notin \Theta_H\}.$$

The problem of testing $\theta \in \Theta_H$ versus $\theta \in \Theta_A$ (that is, $\mu \in M_0$ versus $\mu \notin M_0$) is left invariant by the following transformations:

$$g_{m_0} : (x_{M_0}, x_{M_1}, x_{M^\perp}) \to (x_{M_0} + m_0, x_{M_1}, x_{M^\perp}) \qquad m_0 \in M_0$$
$$g_{\mathcal{O}_{M_1}} : (x_{M_0}, x_{M_1}, x_{M^\perp}) \to (x_{M_0}, \mathcal{O}_{M_1} x_{M_1}, x_{M^\perp}) \qquad \mathcal{O}_{M_1} : M_1 \to M_1$$
$$\text{orthogonal}$$

$$g_{\mathcal{O}_{M^\perp}} \colon (x_{M_0}, x_{M_1}, x_{M^\perp}) \to (x_{M_0}, x_{M_1}, \mathcal{O}_{M^\perp} x_{M^\perp}) \qquad \mathcal{O}_{M^\perp} \colon M^\perp \to M^\perp$$
$$\text{orthogonal}$$
$$g_c \colon (x_{M_0}, x_{M_1}, x_{M^\perp}) \to (c x_{M_0}, c x_{M_1}, c x_{M^\perp}) \qquad\qquad c > 0.$$

Here we have written the typical $x \in V$ as $(x_{M_0}, x_{M_1}, x_{M^\perp})$, where $x_{M_0} = P_{M_0} x$, $x_{M_1} = P_{M_1} x$, and $x_{M^\perp} = P_{M^\perp} x$ are the components of x in the subspaces M_0, $M_1 = M - M_0$, and $M^\perp$, respectively. The general element of the group $\mathcal{G}$ generated by all such transformations can be written in the form

$$g_{m_0, \mathcal{O}_{M_1}, \mathcal{O}_{M^\perp}, c} \colon (x_{M_0}, x_{M_1}, x_{M^\perp}) \to (c x_{M_0} + m_0, c \mathcal{O}_{M_1} x_{M_1}, c \mathcal{O}_{M^\perp} x_{M^\perp}), \qquad (4.4)$$

where $m_0, \mathcal{O}_{M_1}, \mathcal{O}_{M^\perp}$, and c are subject to the above constraints.   •

4.5 Exercise [1]. Let g and h be two transformations of V of the form (4.4). Write f in the form (4.4) for (1) $f = hg$ and (2) $f = g^{-1}$.   ◇

4.6 Exercise [2]. Let $\mathcal{G}$ be a group of transformations leaving the testing problem (4.1) invariant. For $g \in \mathcal{G}$ and $\theta \in \Theta$, let $\bar{g}(\theta) \in \Theta$ be the index such that $P_{\bar{g}(\theta)}$ is the distribution of $g(X)$ when P_θ is the distribution of X, and let $\bar{g}$ be the mapping $\theta \to \bar{g}(\theta)$. Show that: (1) If g is the identity transformation on $\mathfrak{X}$, then $\bar{g}$ is the identity transformation on Θ. (2) $\overline{hg} = \bar{h}\bar{g}$ for all $g, h \in \mathcal{G}$. (3) For each $g \in \mathcal{G}$, $\bar{g}$ is one-to-one, maps Θ_H onto itself, and maps Θ_A onto itself.   ◇

Returning to the general setup, let $\mathcal{G}$ be a group of transformations leaving the problem invariant. Let $g \in \mathcal{G}$ and put $Y = g(X)$. The testing problem based on Y is formally equivalent to that based on X. So if $\phi \colon \mathfrak{X} \to [0, 1]$ is a (possibly randomized) test ($\phi(x)$ being the conditional probability of rejecting H, given that x is the value of X), we could apply it equally well to Y as to X. But Y is distributed according to a hypothesis distribution if and only if X is. To avoid the awkward situation of, for example, deciding that Y is hypothesis distributed ($\phi(Y) = 0$) simultaneously with deciding that X is alternative distributed ($\phi(X) = 1$), we ought to require that we reach the same decision in applying ϕ to X and to Y, that is, that

$$\phi(X) = \phi(Y) = \phi(g(X)).$$

A test ϕ is said to be *invariant under the group* $\mathcal{G}$ if

$$\phi(x) = \phi\big(g(x)\big) \quad \text{for all } x \in \mathfrak{X} \text{ and } g \in \mathcal{G}. \qquad (4.7)$$

The *principle of invariance* states that attention should be limited to invariant tests.

To implement this principle we need a simple way to come up with invariant tests. If x and y are in $\mathfrak{X}$, say $x \sim_\mathcal{G} y$ if there exists a $g \in \mathcal{G}$ such that $g(x) = y$. Since $\mathcal{G}$ is a group, $x \sim_\mathcal{G} y$ is an equivalence relation and $\mathfrak{X}$ splits up into $\sim_\mathcal{G}$ equivalence classes, called *orbits*. ϕ is invariant if and only if it is constant on orbits. Let $\mathbb{M}$ be a mapping from $\mathfrak{X}$ to another space $\mathfrak{M}$. $\mathbb{M}$ is said to be a *maximal invariant* if

   (i) $\mathbb{M}$ is *constant on orbits* ($x \sim_\mathcal{G} y$ implies $\mathbb{M}(x) = \mathbb{M}(y)$), and

   (ii) $\mathbb{M}$ *distinguishes between orbits* ($x \not\sim_\mathcal{G} y$ implies $\mathbb{M}(x) \neq \mathbb{M}(y)$).

Suppose $\mathbb{M}$ is a maximal invariant. It is easily seen that a test $\phi \colon \mathfrak{X} \to [0,1]$ is invariant if and only if there exists a test $\psi \colon \mathfrak{M} \to [0,1]$ such that ϕ is the composition $\psi \mathbb{M}$ of $\mathbb{M}$ and ψ. Hence the invariant tests are the tests that are functions of $\mathbb{M}$.

4.8 Example. Consider the group $\mathcal{G}$ in Example 4.3 above; $\mathcal{G}$ leaves the testing problem for the GLM invariant. We claim that the mapping $\mathbb{M}$ from $\mathfrak{X} = V$ to $\mathfrak{M} \equiv [0,\infty] \cup \{\frac{0}{0}\}$, defined by

$$\mathbb{M}(x) = \frac{\|P_{M_1}x\|^2}{\|Q_M x\|^2} = \frac{\|x_{M_1}\|^2}{\|x_{M^\perp}\|^2} \quad \text{for } x \in V,$$

is a maximal invariant.

$\mathbb{M}$ *is constant on orbits*: Suppose y and x are on the same orbit, so

$$y \equiv (y_{M_0}, y_{M_1}, y_{M^\perp}) = (cx_{M_0} + m_0, c\mathcal{O}_{M_1}x_{M_1}, c\mathcal{O}_{M^\perp}x_{M^\perp})$$

for appropriate m_0, $\mathcal{O}_{M_1}$, $\mathcal{O}_{M^\perp}$, and c. Then $\|y_{M_1}\| = c\|x_{M_1}\|$ and $\|y_{M^\perp}\| = c\|x_{M^\perp}\|$, so that

$$\mathbb{M}(y) = \mathbb{M}(x).$$

$\mathbb{M}$ *distinguishes between orbits*: We need to show that

$$\mathbb{M}(y) = \mathbb{M}(x)$$

implies

$$y \sim_{\mathcal{G}} x.$$

We do this by cases.

Case $1°$: $0 < \mathbb{M}(x) < \infty$ (this is the basic case). Here none of x_{M_1}, $x_{M^\perp}$, y_{M_1}, and $y_{M^\perp}$ are 0. Choose orthogonal transformations $\mathcal{O}_{M_1}$ and $\mathcal{O}_{M^\perp}$ of M_1 and $M^\perp$, respectively, so that

$$\mathcal{O}_{M_1}\left(\frac{x_{M_1}}{\|x_{M_1}\|}\right) = \frac{y_{M_1}}{\|y_{M_1}\|} \quad \text{and} \quad \mathcal{O}_{M^\perp}\left(\frac{x_{M^\perp}}{\|x_{M^\perp}\|}\right) = \frac{y_{M^\perp}}{\|y_{M^\perp}\|}.$$

Then choose c so that

$$\frac{\|y_{M_1}\|}{\|x_{M_1}\|} = c = \frac{\|y_{M^\perp}\|}{\|x_{M^\perp}\|};$$

such a choice is possible because

$$\frac{\|x_{M_1}\|^2}{\|x_{M^\perp}\|^2} = \mathbb{M}(x) = \mathbb{M}(y) = \frac{\|y_{M_1}\|^2}{\|y_{M^\perp}\|^2}.$$

Finally, set $m_0 = y_{M_0} - cx_{M_0}$. Then

$$(y_{M_0}, y_{M_1}, y_{M^\perp}) = (cx_{M_0} + m_0, c\mathcal{O}_{M_1}x_{M_1}, c\mathcal{O}_{M^\perp}x_{M^\perp}).$$

Case 2°: $\mathbb{M}(x) = \frac{0}{0}$. Here $x_{M_1} = 0 = x_{M^\perp}$ and $y_{M_1} = 0 = y_{M^\perp}$, so

$$(y_{M_0}, y_{M_1}, y_{M^\perp}) = (cx_{M_0} + m_0, c\mathcal{O}_{M_1} x_{M_1}, c\mathcal{O}_{M^\perp} x_{M^\perp})$$

for $\mathcal{O}_{M_1} = I_{M_1}$, $\mathcal{O}_{M^\perp} = I_{M^\perp}$, $c = 1$, and $m_0 = y_{M_0} - x_{M_0}$.

4.9 Exercise [2]. Work out the details of

Case 3°: $\mathbb{M}(x) = 0$.

Case 4°: $\mathbb{M}(x) = \infty$.   ◇

Return again to the general setup, where $\mathbb{M}\colon \mathfrak{X} \to \mathfrak{M}$ is assumed to be a maximal invariant for the action of $\mathcal{G}$ on $\mathfrak{X}$. Restricting attention to invariant tests amounts to considering a testing problem on the space $\mathfrak{M}$, as we will now show. For $g \in \mathcal{G}$ and $\theta \in \Theta$, let $\bar{g}(\theta) \in \Theta$ be the index such that

$$\mathrm{Law}_\theta\big(g(X)\big) = \mathrm{Law}_{\bar{g}(\theta)}(X), \tag{4.10}$$

where Law denotes "distribution of." It follows from Exercise 4.6 that the function $\theta \to \bar{g}(\theta)$ is a one-to-one mapping of Θ onto itself and the collection $\bar{\mathcal{G}} = \{\bar{g} : g \in \mathcal{G}\}$ of all such induced transformations of Θ is a group. Let $\mathbb{N}$, mapping Θ *onto* some space Γ, be a maximal invariant for the action of $\bar{\mathcal{G}}$ on Θ; in effect, $\mathbb{N}$ is a means of indexing the orbits of $\bar{\mathcal{G}}$. The basic observation is that the distribution of $\mathbb{M}(X)$ under θ depends on θ only through $\mathbb{N}(\theta)$: For each $g \in \mathcal{G}$,

$$\mathrm{Law}_{\bar{g}(\theta)} \mathbb{M}(X) = \mathrm{Law}_\theta \mathbb{M}\big(g(X)\big) = \mathrm{Law}_\theta \mathbb{M}(X)$$

because $\mathbb{M}$ is invariant. Let

$$\{Q_\gamma : \gamma \in \Gamma\} \tag{4.11$_1$}$$

be the possible distributions of $\mathbb{M}(X)$:

$$Q_\gamma = \mathrm{Law}_\theta \mathbb{M}(X) \quad \text{for any } \theta \text{ such that } \mathbb{N}(\theta) = \gamma. \tag{4.11$_2$}$$

The possible distributions of $\mathbb{M}(X)$ under $H\colon \theta \in \Theta_H$ (respectively, under $A\colon \theta \in \Theta_A$) are $\{Q_\gamma : \gamma \in \Gamma_H\}$ (respectively, $\{Q_\gamma : \gamma \in \Gamma_A\}$), where

$$\Gamma_H = \mathbb{N}(\Theta_H) \quad \text{and} \quad \Gamma_A = \mathbb{N}(\Theta_A); \tag{4.12}$$

note that

$$\Gamma_H \cap \Gamma_A = \varnothing$$

because $\Theta_H \cap \Theta_A = \varnothing$ and Θ_H and Θ_A are each unions of orbits, and

$$\Gamma_H \cup \Gamma_A = \Gamma$$

because $\Theta = \Theta_H \cup \Theta_A$ and $\mathbb{N}$ is onto. Passing from the original problem of testing

$$H\colon \theta \in \Theta_H \quad \text{versus} \quad A\colon \theta \in \Theta_A$$

on the basis of X to the so-called *reduced problem* of testing

$$H: \gamma \in \Gamma_H \quad \text{versus} \quad A: \gamma \in \Gamma_A$$

on the basis of $\mathbb{M}(X)$ is called an *invariance reduction*. In order that the reduced problem be well posed, we need to assume that the distributions Q_γ of (4.11) are distinct.

4.13 Exercise [3]. Let $\mathfrak{X} = \{0, 1\}$, let $\Theta = [0, 1]$, and let P_θ be the distribution on $\mathfrak{X}$ assigning mass θ to 1 and mass $1 - \theta$ to 0. Show that the problem of testing $H: \theta = \frac{1}{2}$ versus $A: \theta \neq \frac{1}{2}$ is left invariant by the transformation of $\mathfrak{X}$ that exchanges 0 and 1. Find maximal invariants for the action of the associated group $\mathcal{G}$ on $\mathfrak{X}$ and the action of $\bar{\mathcal{G}}$ on Θ, and observe that all the Q_γ's are the same. ◇

The idea now is that if a test ψ has an optimal property in the reduced problem, then $\psi\mathbb{M}$ will have the same property among invariant tests in the original problem. The following result illustrates this principle. Recall that the power function β_ϕ of a test ϕ of H versus A based on X is defined by

$$\beta_\phi(\theta) = P_\theta[\text{reject } H \text{ using } \phi] = E_\theta\big(\phi(X)\big) \quad \text{for } \theta \in \Theta. \tag{4.14}$$

One would like $\beta_\phi(\theta)$ to be near 0 for all $\theta \in \Theta_H$ and near 1 for all $\theta \in \Theta_A$. ϕ is said to be *uniformly most powerful of level* α (UMP(α)) in a given class $\mathcal{T}$ of tests if

1° $\phi \in \mathcal{T}$ and $\beta_\phi(\theta) \leq \alpha$ for all $\theta \in \Theta_H$, and

2° for any other test $\tau \in \mathcal{T}$ with $\beta_\tau(\theta) \leq \alpha$ for all $\theta \in \Theta_H$, one has $\beta_\tau(\theta) \leq \beta_\phi(\theta)$ for all $\theta \in \Theta_A$.

Similar definitions apply to tests ψ of H versus A based on $\mathbb{M}(X)$.

4.15 Lemma. *If ψ is a UMP(α) test in the reduced problem, then $\phi \equiv \psi\mathbb{M}$ is UMP(α) among invariant tests in the original problem (in brief, ϕ is UMPI(α)).*

Proof. Suppose ψ is UMP(α) in the reduced problem. Consider $\phi = \psi\mathbb{M}$. This is an invariant test in the original problem. Its power function β_ϕ is related to the power function β_ψ of ψ by the identity

$$\beta_\phi(\theta) = E_\theta\phi(X) = E_\theta\big((\psi\mathbb{M})(X)\big) = E_\theta\big(\psi(\mathbb{M}X)\big) = \beta_\psi\big(\mathbb{N}(\theta)\big) \tag{4.16}$$

for all $\theta \in \Theta$. In particular, $\beta_\phi(\theta) \leq \alpha$ for all $\theta \in \Theta_H$, because $\beta_\psi(\gamma) \leq \alpha$ for all $\gamma \in \Gamma_H$.

Now let ϕ^* be an invariant test in the original problem satisfying $\beta_{\phi^*}(\theta) \leq \alpha$ for all $\theta \in \Theta_H$. Because $\mathbb{M}$ is a maximal invariant, there exists a test ψ^* in the reduced problem such that

$$\phi^* = \psi^*\mathbb{M},$$

and by analogy with (4.16),

$$\beta_{\phi^*}(\theta) = \beta_{\psi^*}\big(\mathbb{N}(\theta)\big) \quad \text{for all } \theta \in \Theta.$$

Consequently, $\beta_{\psi^*}(\gamma) \leq \alpha$ for all $\gamma \in \Gamma_H$. Because ψ is UMP(α) in the reduced problem, we must have

$$\beta_{\psi^*}(\gamma) \leq \beta_\psi(\gamma) \quad \text{for all } \gamma \in \Gamma_A,$$

and this implies that

$$\beta_{\phi^*}(\theta) \leq \beta_\phi(\theta) \quad \text{for all } \theta \in \Theta_A.$$

Hence ϕ is UMPI(α), as claimed.   ∎

4.17 Example. Consider again the testing problem of the GLM. The transformation

$$g \colon (x_{M_0}, x_{M_1}, x_{M^\perp}) \to (c x_{M_0} + m_0, c\mathcal{O}_{M_1} x_{M_1}, c\mathcal{O}_{M^\perp} x_{M^\perp})$$

of $\mathfrak{X}$ induces the transformation

$$\bar{g} \colon (\mu_{M_0}, \mu_{M_1}, \sigma^2) \to (c\mu_{M_0} + m_0, c\mathcal{O}_{M_1}\mu_{M_1}, c^2\sigma^2)$$

of Θ. A maximal invariant for the action of $\bar{\mathcal{G}}$ on Θ is $\mathbb{N} \colon \Theta \to [0, \infty) \equiv \Gamma$, defined by

$$\mathbb{N}(\theta) \equiv \mathbb{N}\big((\mu_{M_0}, \mu_{M_1}, \sigma^2)\big) = \frac{\|\mu_{M_1}\|}{\sigma} = \frac{\|P_{M-M_0}\mu\|}{\sigma} \, .$$

We have seen that a maximal invariant for the action of $\mathcal{G}$ on $\mathfrak{X}$ is

$$\mathbb{M} \colon (x_{M_0}, x_{M_1}, x_{M^\perp}) \to \frac{\|x_{M_1}\|^2/d(M_1)}{\|x_{M^\perp}\|^2/d(M^\perp)} = \frac{\|P_{M-M_0}x\|^2/d(M - M_0)}{\|Q_M x\|^2/d(M^\perp)},$$

and we know that the distribution of $F = \mathbb{M}X$ under $\theta = (\mu_{M_0}, \mu_{M_1}, \sigma^2)$ is

$$\mathcal{F}_{d(M_1), d(M^\perp); \gamma} \equiv Q_\gamma \quad \text{with } \gamma = \mathbb{N}(\theta);$$

this illustrates the general fact that the distribution of the maximal invariant of the data depends on the parameter only through the maximal invariant of the parameter. We have

$$\Gamma_H = \mathbb{N}(\Theta_H) = \{0\} \quad \text{and} \quad \Gamma_A = \mathbb{N}(\Theta_A) = (0, \infty).$$

The reduced problem consists of testing

$$H \colon \gamma \in \Gamma_H, \text{ that is, } \gamma = 0 \quad \text{versus} \quad A \colon \gamma \in \Gamma_A, \text{ that is, } \gamma > 0$$

on the basis of F. Since the family $(Q_\gamma)_{\gamma \geq 0}$ has an increasing MLR in γ by Proposition 3.5, the Neyman-Pearson fundamental lemma (see Lehmann (1959, p. 65)) implies that for any given $0 < \alpha < 1$ there is a UMP(α) test in the reduced problem, namely the test that rejects for $F \geq c$, where c is chosen so that

$$Q_0(\{F \geq c\}) = \alpha.$$

Consequently, the test that rejects for values of X such that

$$\frac{\|P_{M-M_0}X\|^2/d(M-M_0)}{\|Q_M X\|^2/d(M^\perp)} \geq c,$$

that is, the F-test of size α, is UMPI(α) in the original problem. •

To summarize:

4.18 Theorem. *The F-test of size α is uniformly most powerful of level α among tests invariant under the group of transformations (4.4).*

In thinking about this result, it should be borne in mind that transformations that leave the testing problem invariant in the mathematical sense above need not be substantively meaningful. For example, while it might be reasonable to regard a linear combination of the prices of apples, bananas, pears, oranges, and lettuce as an object of interest (the price of fruit salad), it would be nonsensical to include the price of motor oil in the equation.

4.19 Exercise [3]. Suppose the variance parameter σ^2 in the GLM is known; for definiteness, take $\sigma^2 = 1$. The problem of testing $H: \mu \in M_0$ versus $A: \mu \notin M_0$ is left invariant by transformations of the form (4.4) with $c = 1$ and $\mathcal{O}_{M^\perp} = I_{M^\perp}$. Use the approach of this section to find a UMPI(α) test in this situation. ◇

5. Confidence intervals for linear functionals of μ

We return to our old notation:

$$Y \sim N_V(\mu, \sigma^2 I_V), \quad \mu \in M, \quad \sigma^2 > 0. \tag{5.1}$$

Let

$$\psi(\mu) = \langle cv(\psi), \mu \rangle$$

be a nonzero linear functional of μ. Recall that $cv(\psi)$, or $cv\,\psi$ for short, designates the coefficient vector (in M) of ψ. The best unbiased estimator of $\psi(\mu)$ is the GME

$$\hat{\psi}(Y) = \langle cv\,\psi, Y \rangle = \langle cv\,\psi, P_M Y \rangle.$$

The standard deviation of $\hat{\psi}(Y)$ is

$$\sigma_{\hat{\psi}} = \sigma \| cv\,\psi \|;$$

this may be estimated by

$$\hat{\sigma}_{\hat{\psi}} = \hat{\sigma} \| cv\,\psi \|,$$

where

$$\hat{\sigma}^2 = \frac{\|Q_M Y\|^2}{d(M^\perp)}$$

is the best unbiased estimator of σ^2. Because

$$\hat{\psi}(Y) \sim N\left(\psi(\mu), \sigma^2 \| cv\, \psi \|^2\right)$$

independently of

$$\hat{\sigma}^2 \sim \sigma^2 \chi^2_{d(M^\perp)}/d(M^\perp),$$

it follows that

$$\frac{\hat{\psi}(Y) - \psi(\mu)}{\hat{\sigma}_{\hat{\psi}}} = \frac{\left(\hat{\psi}(Y) - \psi(\mu)\right)/(\sigma \|\psi\|)}{\hat{\sigma}/\sigma} \sim t_{d(M^\perp)}, \tag{5.2}$$

as asserted by Exercise 5.1.6. In particular, if $t_m(\beta)$ denotes the upper β fractional point of the t distribution with m degrees of freedom, then

$$\hat{\psi}(Y) \pm t_{d(M^\perp)}\left(\tfrac{\alpha}{2}\right)\hat{\sigma}_{\hat{\psi}}$$

is a $100(1-\alpha)\%$ confidence interval for $\psi(\mu)$:

$$P_{\mu,\sigma^2}\left(\hat{\psi}(Y) \pm t_{d(M^\perp)}\left(\tfrac{\alpha}{2}\right)\hat{\sigma}_{\hat{\psi}} \text{ covers } \psi(\mu)\right) = 1 - \alpha \tag{5.3}$$

for all $\mu \in M$ and all $\sigma^2 > 0$. By Exercise 2.11, this interval fails to cover 0 if and only if the size α F-test of $H\colon \psi(\mu) = 0$ versus $A\colon \psi(\mu) \neq 0$ rejects H.

5.4 Example. Consider simple linear regression: $V = \mathbb{R}^n$, $Y_1, \ldots, Y_n$ are uncorrelated with equal variances σ^2, and $E(Y_i) = \alpha + \beta(x_i - \bar{x})$ for $1 \leq i \leq n$, with $x_1, \ldots, x_n$ known constants. Here

$$\boldsymbol{\mu} = E(\boldsymbol{Y}) = \alpha e + \beta \boldsymbol{v},$$

where $(\boldsymbol{Y})_i = Y_i$, $(e)_i = 1$, and $(\boldsymbol{v})_i = x_i - \bar{x}$ for $1 \leq i \leq n$. Fix an x_0 in $\mathbb{R}$ and consider the point on the population regression line above x_0:

$$\psi_{x_0}(\boldsymbol{\mu}) = \alpha + \beta(x_0 - \bar{x}) = \langle cv\, \psi_{x_0}, \boldsymbol{\mu}\rangle,$$

with
$$cv\, \psi_{x_0} = \frac{e}{\|e\|^2} + (x_0 - \bar{x})\frac{\boldsymbol{v}}{\|\boldsymbol{v}\|^2}.$$

The GME of $\psi_{x_0}(\boldsymbol{\mu})$ is the corresponding point on the fitted line:

$$\hat{\psi}_{x_0}(\boldsymbol{Y}) = \langle cv\, \psi_{x_0}, \boldsymbol{Y}\rangle = \hat{\alpha} + \hat{\beta}(x_0 - \bar{x}),$$

where
$$\hat{\alpha} = \left\langle \frac{e}{\|e\|^2}, \boldsymbol{Y}\right\rangle = \bar{Y} \quad \text{and} \quad \hat{\beta} = \left\langle \frac{\boldsymbol{v}}{\|\boldsymbol{v}\|^2}, \boldsymbol{Y}\right\rangle = \frac{\sum_i (x_i - \bar{x})Y_i}{\sum_i (x_i - \bar{x})^2}$$

are the GMEs of α and β, respectively. One has

$$\sigma_{\hat{\psi}_{x_0}} = \sigma \| cv\, \psi_{x_0} \| = \sigma \sqrt{\frac{1}{\|e\|^2} + \frac{(x_0 - \bar{x})^2}{\|\boldsymbol{v}\|^2}} = \sigma \sqrt{\frac{1}{n} + \frac{(x_0 - \bar{x})^2}{\sum_i (x_i - \bar{x})^2}}$$

and

$$\hat{\sigma}^2 = \frac{\|\boldsymbol{Y} - P_M \boldsymbol{Y}\|^2}{d(M^\perp)} = \frac{\sum_i \left(Y_i - \left(\hat{\alpha} + \hat{\beta}(x_i - \bar{x})\right)\right)^2}{n - 2}.$$

Thus

$$\hat{\alpha} + \hat{\beta}(x_0 - \bar{x}) \pm t_{n-2}\left(\tfrac{\alpha}{2}\right)\hat{\sigma}\sqrt{\frac{1}{n} + \frac{(x_0 - \bar{x})^2}{\sum_i (x_i - \bar{x})^2}} \tag{5.5}$$

is a $100(1 - \alpha)\%$ confidence interval for $\alpha + \beta(x_0 - \bar{x})$.

Up to now x_0 has been fixed. But it is often the case that one wants to estimate $\psi_{x_0}(\boldsymbol{\mu}) = \alpha + \beta(x_0 - \bar{x})$ simultaneously for all, or at least many, values of x_0. The intervals (5.5) are then inappropriate for $100(1 - \alpha)\%$ confidence, because

$$P_{\boldsymbol{\mu},\sigma^2}\left(\hat{\psi}_{x_0}(Y) \pm t_{n-2}\left(\tfrac{\alpha}{2}\right)\hat{\sigma}_{\hat{\psi}_{x_0}} \text{ covers } \psi_{x_0}(\boldsymbol{\mu}) \text{ for all } x_0 \in \mathbb{R}\right) < 1 - \alpha \tag{5.6}$$

(see Exercise 5.20 below). We will develop a method of making simultaneous inferences on $\psi_{x_0}(\boldsymbol{\mu})$ for all x_0 — in fact, we will treat the general problem of making simultaneous inferences on $\psi(\boldsymbol{\mu})$ for an arbitrary family of linear functionals in the context of the GLM. •

Return to the setting of (5.1). Let $\mathcal{K}$ be a collection of linear functionals of $\boldsymbol{\mu}$ and set

$$K = \{ cv\,\psi : \psi \in \mathcal{K} \} \subset M. \tag{5.7}$$

Let $\mathcal{L}$ be the subspace generated by $\mathcal{K}$ in the vector space M° of all linear functionals on M and set

$$L = \{ cv\,\psi : \psi \in \mathcal{L} \} \subset M; \tag{5.8}$$

equivalently,

$$L = \text{span}(K).$$

$\mathcal{L}$ and L are isomorphic, so

$$d(\mathcal{L}) = d(L).$$

Assume some $\psi \in \mathcal{K}$ is nonzero, so $d(\mathcal{L}) \geq 1$. We will produce a constant C (depending on $d(L)$, $d(M^\perp)$, and α) such that

$$P_{\boldsymbol{\mu},\sigma^2}\left(\hat{\psi}(Y) \pm C\hat{\sigma}_{\hat{\psi}} \text{ covers } \psi(\boldsymbol{\mu}) \text{ for all } \psi \in \mathcal{L}\right) = 1 - \alpha \tag{5.9}$$

for all $\boldsymbol{\mu}, \sigma^2$; this says that the intervals

$$\hat{\psi}(Y) \pm C\hat{\sigma}_{\hat{\psi}} \tag{5.10}$$

cover the $\psi(\boldsymbol{\mu})$'s for $\psi \in \mathcal{L}$ with *simultaneous* confidence $100(1 - \alpha)\%$. Of course, (5.9) implies

$$P_{\boldsymbol{\mu},\sigma^2}\left(\hat{\psi}(Y) \pm C\hat{\sigma}_{\hat{\psi}} \text{ covers } \psi(\boldsymbol{\mu}) \text{ for all } \psi \in \mathcal{K}\right) \geq 1 - \alpha \tag{5.11}$$

for all μ, σ^2, whence the intervals (5.10) cover the $\psi(\mu)$'s for $\psi \in \mathcal{K}$ with simultaneous confidence *at least* $100(1 - \alpha)\%$.

Now

$$\hat{\psi}(Y) \pm C\hat{\sigma}_{\hat{\psi}} \quad \text{covers} \quad \psi(\mu) \quad \text{for all } \psi \in \mathcal{L}$$

if and only if

$$\sup\nolimits_{0 \neq \psi \in \mathcal{L}} \frac{\left(\hat{\psi}(Y) - \psi(\mu)\right)^2}{\hat{\sigma}_{\hat{\psi}}^2} \leq C^2.$$

But

$$\sup\nolimits_{0 \neq \psi \in \mathcal{L}} \frac{\left(\hat{\psi}(Y) - \psi(\mu)\right)^2}{\hat{\sigma}_{\hat{\psi}}^2} = \sup\nolimits_{0 \neq \psi \in \mathcal{L}} \frac{\left(\langle cv\,\psi, Y - \mu\rangle\right)^2}{\|\, cv\,\psi\|^2\, \hat{\sigma}^2}$$

$$= \frac{1}{\hat{\sigma}^2} \sup\nolimits_{0 \neq x \in L} \frac{\left(\langle x, Y - \mu\rangle\right)^2}{\|x\|^2} = \frac{1}{\hat{\sigma}^2} \sup\nolimits_{0 \neq x \in L} \left(\left\langle \frac{x}{\|x\|}, P_L(Y - \mu)\right\rangle\right)^2$$

$$= \frac{\|P_L(Y - \mu)\|^2}{\hat{\sigma}^2} \equiv Q, \tag{5.12}$$

the last equality holding by the Cauchy-Schwarz inequality (2.1.21). Since $Y - \mu \sim N_V(0, \sigma^2 I_V)$ and $L \perp M^\perp$, we have

$$\frac{Q}{d(L)} = \frac{\|P_L(Y - \mu)\|^2/d(L)}{\|Q_M(Y - \mu)\|^2/d(M^\perp)} \sim \mathcal{F}_{d(L), d(M^\perp)};$$

notice that it is the central $\mathcal{F}$ distribution that figures here. It follows that for all μ, σ^2,

$$P_{\mu,\sigma^2}\left(\hat{\psi}(Y) \pm C\hat{\sigma}_{\hat{\psi}} \text{ covers } \psi(\mu) \text{ for all } \psi \in \mathcal{L}\right)$$
$$= P_{\mu,\sigma^2}\left(Q \leq C^2\right) = \mathcal{F}_{d(L), d(M^\perp)}\left(\left[0, C^2/d(L)\right]\right). \tag{5.13}$$

Letting $\mathcal{F}_{f_1, f_2}(\alpha)$ denote the upper α fractional point of the $\mathcal{F}$ distribution with f_1 and f_2 degrees of freedom, and setting

$$S_{f_1, f_2}(\alpha) = \sqrt{f_1 \mathcal{F}_{f_1, f_2}(\alpha)}, \tag{5.14}$$

we find that (5.9) holds with $C = S_{d(L), d(M^\perp)}(\alpha)$:

5.15 Theorem. *If $\mathcal{L}$ is a subspace of M^o and $L = \{cv\,\psi : \psi \in \mathcal{L}\} \subset M$ is the corresponding subspace of coefficient vectors, then the intervals*

$$\hat{\psi}(Y) \pm S_{d(L), d(M^\perp)}(\alpha)\hat{\sigma}_{\hat{\psi}} \tag{5.16}$$

cover the $\psi(\mu)$'s for $\psi \in \mathcal{L}$ with simultaneous confidence $100(1 - \alpha)\%$.

The intervals (5.16) are called *Scheffé* intervals; $S_{d(L),d(M^\perp)}(\alpha)$ is the *Scheffé multiplier*. Note that $S_{1,f}(\alpha) = t_f\left(\frac{\alpha}{2}\right)$, so that when $\mathcal{L} = [\psi]$ is one-dimensional, Theorem 5.15 reduces to the simple assertion (5.3) that $\hat{\psi}(Y) \pm t_{d(M^\perp)}\left(\frac{\alpha}{2}\right)\hat{\sigma}_{\hat{\psi}}$ is a $100(1-\alpha)\%$ confidence interval for $\psi(\mu)$.

5.17 Example. Return to the simple linear regression model. Put

$$\mathcal{K} = \{\,\psi_{x_0} : x_0 \in \mathbb{R}\,\},$$

so

$$K = \left\{\,\frac{e}{\|e\|^2} + (x_0 - \bar{x})\frac{v}{\|v\|^2} : x_0 \in \mathbb{R}\,\right\}$$

and

$$L = \operatorname{span}(K) = [e, v] = M \quad \text{and} \quad d(L) = d(M) = 2.$$

From Theorem 5.15, the Scheffé intervals

$$\hat{\psi}_{x_0}(Y) \pm S_{2,n-2}(\alpha)\hat{\sigma}_{\hat{\psi}_{x_0}} \tag{5.18}$$

cover the various $\psi_{x_0}(\mu) = \alpha + \beta(x_0 - \bar{x})$ for $x_0 \in \mathbb{R}$ with simultaneous confidence *at least* $100(1-\alpha)\%$; note that $\mathcal{K}$ is properly contained in $\mathcal{L} = \operatorname{span}(\mathcal{K})$.　　　　●

5.19 Exercise [2]. Let $\mathcal{K}$, K, $\mathcal{L}$, and L be as in the discussion leading up to Theorem 5.15. Show that for any constant C, simultaneous coverage by the intervals $\hat{\psi}(Y) \pm C\hat{\sigma}_{\hat{\psi}}$ for all $\psi \in \mathcal{K}$ is equivalent to simultaneous coverage for all $\psi \in \mathcal{L}$ provided

$$L = \bigcup_{x \in K} \operatorname{span}(x).$$

　　　◇

5.20 Exercise [3]. Use the result of the preceding exercise to deduce that the intervals (5.18) in fact have simultaneous confidence *exactly* $100(1-\alpha)\%$. Use (5.13) to evaluate the left-hand side of (5.6). Verify that for large n the probability of simultaneous coverage for all $x_0 \in \mathbb{R}$ by the t-intervals (5.5) is given approximately by the formula $1 - \exp(-\frac{1}{2}t_\infty^2(\frac{\alpha}{2}))$, as exemplified by the following table:

Nominal Confidence Level	Actual Simultaneous Confidence Level	Nominal Failure Rate	Actual Failure Rate
90.0%	74.0%	10.0%	26.0%
95.0%	85.0%	5.0%	15.0%
99.0%	96.4%	1.0%	3.6%

　　　◇

5.21 Exercise [5]. Suppose the variance parameter σ^2 in (0.1) is known. Let $\mathcal{K} \subset M^\circ$ be the line segment $\{\,w\psi_0 + (1-w)\psi_1 : 0 \le w \le 1\,\}$ between two given linear functionals ψ_0 and ψ_1 of μ, and suppose $0 \notin \mathcal{K}$. Let

$$\theta = \frac{1}{2}\arccos\left(\frac{\langle\psi_0,\psi_1\rangle}{\|\psi_0\|\,\|\psi_1\|}\right) = \frac{1}{2}\arccos\left(\frac{\langle cv\,\psi_0, cv\,\psi_1\rangle}{\|cv\,\psi_0\|\,\|cv\,\psi_1\|}\right) \tag{5.22}$$

be half the angle between ψ_0 and ψ_1; by convention, $0 \le \theta \le \frac{1}{2}\pi$. Show that the failure probability

$$q_{\theta,C} \equiv P_\mu\left(\hat{\psi}(Y) \pm C\sigma_{\hat{\psi}} \text{ fails to cover } \psi(\mu) \text{ for at least one } \psi \in \mathcal{K}\right) \tag{5.23}$$

depends only on θ and C and is given by the formula

$$q_{\theta,C} = \frac{2}{\pi}\left(\theta e^{-C^2/2} + \int_0^{\pi/2-\theta} e^{-C^2/(2\cos^2(a))}\,da\right). \tag{5.24}$$

[Hint: Use the ideas behind (5.12).]   ◇

5.25 Figure. *Failure probabilities and confidence coefficients for simultaneous confidence intervals for $\psi(\mu)$ for $\psi \in \mathcal{K}$ in the case when $\mathcal{K}$ is a line segment in $M°$. Let $\mathcal{K}$ be the collection of linear functionals of μ described in Exercise 5.21 and let $q_{\theta,C}$ be the failure probability defined by (5.23)–(5.24). In the left graph below, $q_{\theta,C}$ is plotted versus $\sin(\theta)$ for $C = t_\infty(\frac{\alpha}{2}) \equiv z(\frac{\alpha}{2})$ with $\alpha = 0.10$, 0.05, and 0.01. The graph shows that $q_{\theta,z(\alpha/2)}$ is substantially larger than $\alpha = q_{0,z(\alpha/2)}$ unless the angle θ is quite small. The right graph below exhibits the confidence coefficient $C = C_\theta(\alpha)$ that makes $q_{\theta,C} = \alpha$. The quantity $D_\theta(\alpha) = \alpha \exp(\frac{1}{2}C_\theta^2(\alpha))$ is plotted on the vertical axis against $\sin(\theta)$ on the horizontal axis, for $\alpha = 0.10$, 0.05, and 0.01. The graph shows that for a given α, $D_\theta(\alpha)$ is to a close approximation given by linear interpolation with respect to $\sin(\theta)$ between $D_0(\alpha) = \alpha \exp(\frac{1}{2}z^2(\frac{\alpha}{2}))$ and $D_{\pi/2}(\alpha) = 1$. $C_{\pi/2}(\alpha)$ is the Scheffé multiplier $S_{2,\infty}(\alpha) = \sqrt{2\log(1/\alpha)}$.*

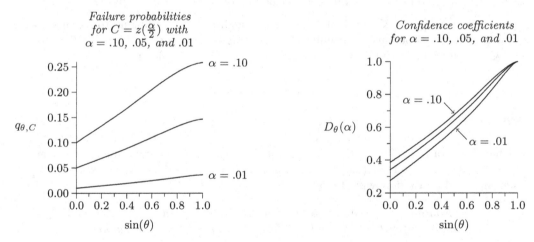

5.26 Exercise [3]. As in Example 5.4, let $\hat{\psi}_{x_0}(\boldsymbol{Y})$ be the GME of $\psi_{x_0}(\boldsymbol{\mu}) = \alpha + \beta(x_0 - \bar{x})$ in the simple linear regression model. Suppose that σ is known. Using the result of the preceding exercise, show that for $\xi \geq 0$ and $C \geq 0$,

$$P_{\boldsymbol{\mu}}\left(\hat{\psi}_{x_0}(\boldsymbol{Y}) \pm C\sigma_{\hat{\psi}_{x_0}} \text{ fails to cover } \psi_{x_0}(\boldsymbol{\mu}) \text{ for some } x_0 \text{ with } |x_0 - \bar{x}| \leq \xi\right) = q_{\theta,C}, \tag{5.27}$$

where $q_{\theta,C}$ is given by (5.24) with

$$\theta = \arctan\left(\frac{\xi}{\sqrt{\frac{1}{n}\sum_{1\leq i\leq n}(x_i - \bar{x})^2}}\right). \tag{5.29}$$

Verify that $\theta = \frac{1}{4}\pi$ if half of the design points $x_1,\ldots,x_n$ are at ξ and the other half are at $-\xi$, and that $\theta \approx \frac{1}{3}\pi$ if the x_i's are evenly spaced over the interval from $-\xi$ to ξ.   ◇

The Scheffé intervals (5.16) have an interesting and important relationship to the F-test for

$$H: \psi(\mu) = 0 \text{ for all } \psi \in \mathcal{K} \text{ versus } A: \psi(\mu) \neq 0 \text{ for some } \psi \in \mathcal{K}. \qquad (5.30)$$

This is a testing problem of the usual kind, because

$$\begin{aligned} M_0 &\equiv \{\, \mu \in M : \psi(\mu) = 0 \text{ for all } \psi \in \mathcal{K} \,\} \\ &= \{\, \mu \in M : \psi(\mu) = 0 \text{ for all } \psi \in \mathcal{L} \,\} \\ &= \{\, \mu \in M : \langle x, \mu \rangle = 0 \text{ for all } x \in L \,\} \\ &= M - L \end{aligned} \qquad (5.31)$$

is indeed a subspace of M; note that

$$M - M_0 = L = \{\, cv\,\psi : \psi \in \mathcal{L} \,\}.$$

The hypothesis $\mu \in M_0$ fails if $\psi(\mu) \neq 0$ for some $\psi \in \mathcal{L}$. The canonical estimator of $\psi(\mu)$ is the GME $\hat{\psi}(Y)$; say that $\hat{\psi}(Y)$ is *significantly different from zero* (SDFZ) *at level* α according to the Scheffé criterion if

$$\hat{\psi}(Y) \pm S_{d(L),d(M^\perp)}(\alpha)\hat{\sigma}_{\hat{\psi}} \text{ does not cover } 0$$

$$\iff |\hat{\psi}(Y)| \geq S_{d(L),d(M^\perp)}(\alpha)\hat{\sigma}_{\hat{\psi}}. \qquad (5.32)$$

One then has

$$\hat{\psi}(Y) \text{ is SDFZ at level } \alpha \text{ for some } \psi \in \mathcal{L}$$

$$\iff \sup_{\psi \in \mathcal{L}} |\hat{\psi}(Y)|^2 / \hat{\sigma}_{\hat{\psi}}^2 \geq d(L) \mathcal{F}_{d(L),d(M^\perp)}(\alpha)$$

$$\iff \frac{\|P_L(Y)\|^2 / d(L)}{\|Q_M(Y)\|^2 / d(M^\perp)} \geq \mathcal{F}_{d(M - M_0),d(M^\perp)}(\alpha)$$

$$\iff \text{the size } \alpha \ F\text{-test rejects the hypothesis } H: \mu \in M_0, \qquad (5.33)$$

where the second $\iff$ uses (5.12) with $\mu = 0$. For the case $d(L) = 1$, this phenomenon was pointed out in the discussion following (5.2).

5.34 Exercise [1]. Show that the testing problem (0.2) can always be put in the form (5.30) for a suitable choice of $\mathcal{K}$. ◇

5.35 Exercise [2]. Let $\mathcal{L}$ be a subspace of the vector space of linear functionals on M and let M_0 be defined by (5.30) with $\mathcal{K} = \mathcal{L}$. Write $\mu \in M$ as (μ_{M_0}, μ_{M_1}), where $\mu_{M_0} = P_{M_0}\mu$ and $\mu_{M_1} = P_{M_1}\mu$ with $M_1 = M - M_0$. For each $m_1 \in M_1$, let $\mathfrak{A}_{m_1}$ be the acceptance region of a nonrandomized test of the hypothesis $H_{m_1} : \mu_{M_1} = m_1$ versus the alternative $A_{m_1} : \mu_{M_1} \neq m_1$. For each $v \in V$, let $\mathfrak{C}(v)$ be the subset of M_1 defined by the rule

$$m_1 \in \mathfrak{C}(v) \iff v \in \mathfrak{A}_{m_1};$$

$\mathfrak{C}(Y)$ can be thought of as the set of values of μ_{M_1} that are plausible in the light of the data Y, as judged by the given family of tests. (1) Show that each acceptance region $\mathfrak{A}_{m_1}$ is *similar of size* α, in the sense that

$$P_{\mu_{M_0},m_1,\sigma^2}(Y \notin \mathfrak{A}_{m_1}) = \alpha$$

for all $\mu_{M_0} \in M_0$ and $\sigma^2 > 0$, if and only if $\mathfrak{C}$ is a $100(1-\alpha)\%$ *confidence region procedure for* μ_{M_1}, in the sense that

$$P_{\mu_{M_0},\mu_{M_1},\sigma^2}(\mathfrak{C}(Y) \text{ covers } \mu_{M_1}) = 1 - \alpha$$

for all $\mu_{M_0} \in M_0$, $\mu_{M_1} \in M_1$, and $\sigma^2 > 0$. (2) Show that when $\mathfrak{A}_{m_1}$ is taken as the acceptance region of the size α F-test of $H: \mu \in m_1 + M_0$ versus $A: \mu \notin m_1 + M_0$, then

$$\mathfrak{C}(Y) = \{\, m_1 \in M_1 : \|m_1 - P_{M_1}Y\| \le S_{d(M_1),d(M^\perp)}(\alpha)\hat\sigma \,\};$$

moreover, for each $\mu_{M_1} \in M_1$,

$$Y \in \mathfrak{A}_{\mu_{M_1}} \iff \mathfrak{C}(Y) \text{ covers } \mu_{M_1} \iff \begin{bmatrix} \hat\psi(Y) \pm S_{d(M_1),d(M^\perp)}(\alpha)\hat\sigma_{\hat\psi} \\ \text{covers } \psi(\mu_{M_1}) \text{ for all } \psi \in \mathcal{L} \end{bmatrix}. \qquad \diamond$$

5.36 Example. Consider one-way ANOVA, in which $V = \{(x_{ij})_{j=1,\ldots,n_i;\, i=1,\ldots,p} \in \mathbb{R}^n\}$ with $n = \sum_i n_i$ and $E(Y_{ij}) = \beta_i$ for $j = 1,\ldots,n_i$, $i = 1,\ldots,p$, so

$$M = \Big\{ \sum_{1 \le i \le p} \boldsymbol{v}^{(i)}\beta_i : \beta_1,\ldots,\beta_p \in \mathbb{R} \Big\}$$

with $(\boldsymbol{v}^{(i)})_{i'j'} = \delta_{ii'}$. Notice that

$$\beta_i = \Big\langle \frac{\boldsymbol{v}^{(i)}}{\|\boldsymbol{v}^{(i)}\|^2}, \mu \Big\rangle \equiv \psi_i(\mu) \quad \text{for } 1 \le i \le p.$$

A linear functional of μ of the form

$$\psi(\mu) = \sum_{1 \le i \le p} c_i \psi_i(\mu) = \sum_{1 \le i \le p} c_i \beta_i \tag{5.37_1}$$

with

$$\sum_{1 \le i \le p} c_i = 0 \tag{5.37_2}$$

is called a *contrast* in the group means $\beta_1,\ldots,\beta_p$. For example, $\beta_2 - \beta_1$ and $\beta_1 - \frac{1}{2}(\beta_2 + \beta_3)$ are contrasts, but β_1 is not. The space $\mathcal{C}$ of all contrasts has dimension $p - 1$, because it is the range of the nonsingular linear transformation

$$c \to \sum_{1 \le i \le p} c_i \psi_i$$

defined on the $p - 1$ dimensional subspace

$$\Big\{ c \in \mathbb{R}^p : \sum_{1 \le i \le p} c_i = 0 \Big\} = \mathbb{R}^p - [e]$$

of $\mathbb{R}^p$. The GME of the contrast $\sum_{1 \leq i \leq p} c_i \beta_i$ is

$$\hat{\psi}(\boldsymbol{Y}) = \sum_{1 \leq i \leq p} c_i \hat{\beta}_i = \sum_{1 \leq i \leq p} c_i \bar{Y}_i;$$

it has estimated standard deviation

$$\hat{\sigma}_{\hat{\psi}} = \hat{\sigma} \sqrt{\sum_{1 \leq i \leq p} c_i^2 / n_i}$$

with

$$\hat{\sigma}^2 = \frac{\|\boldsymbol{Y} - P_M \boldsymbol{Y}\|^2}{d(M^\perp)} = \frac{\sum_{1 \leq i \leq p} \sum_{1 \leq j \leq n_i} (Y_{ij} - \bar{Y}_i)^2}{n - p}.$$

By Theorem 5.15,

$$P_{\mu, \sigma^2} \left(\hat{\psi}(\boldsymbol{Y}) \pm S_{p-1, n-p}(\alpha) \hat{\sigma}_{\hat{\psi}} \text{ covers } \psi(\mu) \text{ for all contrasts } \psi \right) = 1 - \alpha$$

for all $\mu \in M$ and $\sigma^2 > 0$. Moreover, by (5.33), some estimated contrast $\hat{\psi}(\boldsymbol{Y})$ is significantly different from 0 at level α if and only if the size α F-test of the hypothesis $H: \psi(\mu) = 0$ for all $\psi \in \mathcal{C}$, that is, of the hypothesis

$$H: \beta_1 = \beta_2 = \cdots = \beta_p$$

rejects:

$$\frac{\|P_{M-M_0} \boldsymbol{Y}\|^2 / (p-1)}{\|Q_M \boldsymbol{Y}\|^2 / (n-p)} = \frac{\sum_i n_i (\bar{Y}_i - \bar{Y})^2 / (p-1)}{\sum_i \sum_j (Y_{ij} - \bar{Y}_i)^2 / (n-p)} \geq \mathcal{F}_{p-1, n-p}(\alpha). \qquad \bullet$$

5.38 Exercise [2]. In the context of the GLM, suppose $\psi_1, \ldots, \psi_p$ are p linear functionals of μ. Show that the space

$$\mathcal{C} = \left\{ \sum_{1 \leq i \leq p} c_i \psi_i : \sum_{1 \leq i \leq p} c_i = 0 \right\}$$

of contrasts in the ψ_i's can be written in the form

$$\mathcal{C} = \operatorname{span}(\psi_i - \bar{\psi} : 1 \leq i \leq p),$$

where $\bar{\psi} = \sum_{1 \leq i \leq p} w_i \psi_i / w$ is the weighted average of the ψ_i's with respect to any given nonnegative weights $w_1, \ldots, w_p$ with sum $w > 0$. ◇

5.39 Exercise [3]. Suppose $\psi_1, \ldots, \psi_p$ are p nonzero orthogonal linear functionals of μ. Put

$$M_0 = \{ \mu \in M : \psi(\mu) = 0 \text{ for all contrasts } \psi \text{ in the } \psi_i\text{'s} \}.$$

Show that

$$d(M - M_0) = p - 1$$

and

$$\|P_{M-M_0} \boldsymbol{Y}\|^2 = \sum_{1 \leq i \leq p} w_i (\hat{\psi}_i(\boldsymbol{Y}) - \bar{\hat{\psi}}(\boldsymbol{Y}))^2 = \sum_{1 \leq i \leq p} w_i \hat{\psi}_i^2(\boldsymbol{Y}) - w \bar{\hat{\psi}}^2(\boldsymbol{Y}), \qquad (5.40)$$

where

$$\hat{\psi}_i(Y) \text{ is the GME of } \psi_i(\mu)$$
$$\frac{1}{w_i} = \text{Var}(\hat{\psi}_i(Y))/\sigma^2 = \|\psi_i\|^2 \tag{5.41}$$

for $1 \le i \le p$ and

$$\hat{\bar{\psi}}(Y) \text{ is the GME of } \bar{\psi}(\mu) = \frac{1}{w} \sum_{1 \le i \le p} w_i \psi_i(\mu)$$
$$\frac{1}{w} = \frac{1}{\sum_{1 \le i \le p} w_i} = \text{Var}(\hat{\bar{\psi}})/\sigma^2 = \|\bar{\psi}\|^2. \tag{5.42}$$

[Hint: Check that $[\psi_1 - \bar{\psi}, \dots, \psi_p - \bar{\psi}] = [\psi_1, \dots, \psi_p] - [\bar{\psi}]$ and use the result of Exercise 5.38.]   ◇

5.43 Exercise [2]. Use the result of the preceding exercise to obtain the numerator of the F-statistic in Example 5.36.   ◇

6. Problem set: Wald's theorem

This problem set is devoted to the proof of yet another optimality property of the F-test. Consider the testing problem (0.2): For given subspaces M and M_0 of V, we have

$$Y \sim P_\theta \equiv N_V(\mu, \sigma^2 I_V)$$

for some unknown $\theta \equiv (\mu, \sigma^2)$ in

$$\Theta = M \times (0, \infty),$$

and we wish to test

$$H: \theta \in \Theta_H \quad \text{versus} \quad A: \theta \in \Theta_A, \tag{6.1}$$

where

$$\Theta_H = M_0 \times (0, \infty) \quad \text{and} \quad \Theta_A = \{\theta \in \Theta : \theta \notin \Theta_H\}.$$

We will sometimes write the generic $\theta \in \Theta$ as $(\mu_{M_0}, \mu_{M_1}, \sigma^2)$, where $\mu_{M_0} = P_{M_0}\mu$ and $\mu_{M_1} = P_{M_1}\mu$ are the components of $\mu \in M$ in M_0 and $M_1 = M - M_0$, respectively. A (possibly) *randomized test* ϕ of H versus A is a mapping from V to $[0, 1]$ with the interpretation that for each $v \in V$, $\phi(v)$ is the conditional probability of rejecting H when $Y = v$. The power function of ϕ is the mapping β_ϕ from Θ to $[0, 1]$ defined by

$$\beta_\phi(\theta) = E_\theta \phi(Y) = \int_V \phi(y) P_\theta(dy) = P_\theta\big(\text{reject } H \text{ using } \phi\big).$$

For given $\mu_{M_0} \in M_0$, $\rho > 0$, and $\sigma^2 > 0$, let

$$\bar{\beta}_\phi(\mu_{M_0}, \rho, \sigma^2) \tag{6.2}$$

denote the average of $\beta_\phi(\theta)$ with respect to alternatives $\theta = (\mu_{M_0}, \mu_{M_1}, \sigma^2)$ with μ_{M_1} chosen at random according to the uniform distribution U_ρ on the sphere $\{ m \in M_1 : \|m\| = \rho \}$. This problem set is devoted to proving

6.3 Theorem (Wald's theorem). *In regard to the problem* (6.1), *among tests* ϕ *such that*

$$\beta_\phi(\theta) = \alpha \quad \text{for all } \theta \in \Theta_H, \tag{6.4}$$

the size α F-*test maximizes the average power* $\bar{\beta}_\phi(\mu_{M_0}, \rho, \sigma^2)$ *for every* $\mu_{M_0} \in M_0$, $\rho > 0$, *and* $\sigma^2 > 0$.

Loosely speaking, the F-test is best when whatever might be $\mu_{M_0} \in M_0$ and $\sigma^2 > 0$, all alternatives with μ_{M_1} of a given magnitude are of equal, and hence "uniform," concern. A test ϕ such that (6.4) holds is said to be *similar of size* α. A test ϕ such that

$$\beta_\phi(\theta) \le \alpha \text{ for all } \theta \in \Theta_H \quad \text{and} \quad \beta_\phi(\theta) \ge \alpha \text{ for all } \theta \in \Theta_A$$

is said to be *unbiased of level* α.

A. Show that the size α F-test is unbiased of level α and (for the problem at hand) every unbiased test of level α is similar of size α.
[Hint: The second assertion follows from the continuity of $\beta_\phi(\theta)$ in θ (see Exercise 5.2.4).]   ○

A reformulation of Wald's theorem is somewhat easier to prove. Put

$$\Theta^* = \{ (\mu_{M_0}, \rho, \sigma^2) : \mu_{M_0} \in M_0, \ \rho \ge 0, \ \sigma^2 > 0 \} = M_0 \times [0, \infty) \times (0, \infty),$$

and for $\theta^* \in \Theta^*$ let $P_{\theta^*}^*$ be the average of $P_{(\mu_{M_0}, \mu_{M_1}, \sigma^2)}$ with μ_{M_1} chosen at random according to the uniform distribution U_ρ above. Consider testing

$$H^* : \theta^* \in \Theta_H^* \quad \text{versus} \quad A^* : \theta^* \in \Theta_A^*, \tag{6.5}$$

where

$$\Theta_H^* = \{ (\mu_{M_0}, \rho, \sigma^2) \in \Theta^* : \rho = 0 \} \quad \text{and} \quad \Theta_A^* = \{ \theta^* \in \Theta^* : \theta^* \notin \Theta_H^* \}.$$

For any test $\phi : V \to [0, 1]$, write β_ϕ^* for the power function of ϕ in the problem (6.5):

$$\beta_\phi^*(\theta^*) = \int_V \phi(y) \, P_{\theta^*}^*(dy) = P_{\theta^*}^* \big(\text{reject } H^* \text{ using } \phi \big).$$

B. Show that Wald's theorem is equivalent to

6.6 Theorem. *In regard to problem* (6.5), *the size* α F-*test is uniformly most powerful among tests* ϕ *satisfying the similarity condition*

$$\beta_\phi^*(\theta^*) = \alpha \quad \text{for all } \theta^* \in \Theta_H^*. \tag{6.7} \ \circ$$

Parts **C** through **G** deal with the proof of Theorem 6.6. To begin with, here is a technical result that is called upon in the later stages of the proof.

C. Let U be a random variable uniformly distributed over the unit sphere $\{\, m_1 \in M_1 : \|m\|_1 = 1 \,\}$ in M_1. Show that the function

$$G(c) \equiv E(e^{c\langle m_1, U\rangle}) \tag{6.8}$$

does not depend on the unit vector m_1 in M_1 and is continuous and strictly increasing in $c \geq 0$.

[Hint: For the first assertion, use the fact that U and $\mathcal{O}U$ have the same distributions for all orthogonal transformations $\mathcal{O}$ of M_1 into itself. For the second assertion, look at the first two derivatives of G.]   ∘

The next part examines the distributions $P_{\theta^*}^*$ for $\theta^* \in \Theta^*$.

D. Show that the density of $P_{\theta^*}^*$ (with respect to Lebesgue measure on V) is

$$f_{\theta^*}^*(y) \equiv \left(\frac{1}{2\pi\sigma^2}\right)^{n/2} \exp\left(-\frac{\|\mu_{M_0}\|^2 + \rho^2}{2\sigma^2}\right) \exp\left(\langle P_{M_0}y, \frac{\mu_{M_0}}{\sigma^2}\rangle\right)$$
$$\times\, G\left(\|P_{M_1}y\| \frac{\rho}{\sigma^2}\right) \exp\left(-\frac{\|y\|^2}{2\sigma^2}\right),$$

where $\theta^* = \left(\mu_{M_0}, \rho, \sigma^2\right)$, $n = \dim(V)$, and G is defined by (6.8). Deduce that the triple

$$T_0 = P_{M_0}Y, \quad T_1 = \|P_{M_1}Y\|, \quad \text{and} \quad T_2 = \|Y\|^2$$

is a sufficient statistic for the family $\{\, P_{\theta^*}^* : \theta^* \in \Theta^* \,\}$.

[Hint: For the second assertion, use the factorization criterion for sufficiency (see Lehmann (1959, p. 49)).]   ∘

In view of the sufficiency of T_0, T_1, and T_2, we may limit attention to tests ϕ of the form

$$\phi(y) = \psi\left(P_{M_0}y, \|P_{M_1}y\|, \|y\|^2\right), \tag{6.9}$$

where $\psi : M_0 \times [0, \infty) \times [0, \infty) \to [0, 1]$. Bear in mind that we are trying to test whether ρ is 0 or, equivalently, whether

$$\eta \equiv \frac{\rho}{\sigma^2} \tag{6.10}$$

is 0, regardless of the values of the nuisance parameters μ_{M_0}/σ^2 and $1/\sigma^2$ figuring in $f_{\theta^*}^*(y)$. The conditioning argument of the next part makes these nuisance parameters "disappear."

E. Show that for each $\tau_0 \in M_0$ and $\tau_2 \geq 0$, the conditional $P_{\theta^*}^*$-distribution of T_1 given $T_0 = \tau_0$ and $T_2 = \tau_2$ depends on θ^* only through the parameter η of (6.10); specifically, for $\tau_1 \geq 0$, the $P_{\theta^*}^*$-conditional density $f_{\theta^*}^*(\tau_1 \mid \tau_0, \tau_2)$ of T_1 given $T_0 = \tau_0$ and $T_2 = \tau_2$ has the form

$$f_{\theta^*}^*(\tau_1 \mid \tau_0, \tau_2) = G(\tau_1\eta)\, h_{\tau_0, \tau_2}(\tau_1)\, C_{\tau_0, \tau_2}(\eta),$$

where G is as in (6.8) and $h_{\tau_0,\tau_2}(\tau_1)$ and $C_{\tau_0,\tau_2}(\eta)$ depend only on the arguments shown.

[Hint: First work out the form of the $P_{\theta^*}^*$-joint density of T_0, T_1, and T_2, and then use the standard formula for a conditional density; it is not necessary to evaluate $f_{\theta^*}^*(\tau_1 \mid \tau_0, \tau_2)$ explicitly.]   ○

The above result suggests testing (6.5) conditionally on T_0 and T_2. Write $Q_{\eta|\tau_0,\tau_2}$ for the conditional $P_{\theta^*}^*$-distribution of T_1, given $T_0 = \tau_0$ and $T_2 = \tau_2$.

F. Show that in regard to the family $\{\, Q_{\eta|\tau_0,\tau_2} : \eta \geq 0 \,\}$ of distributions on $[0,\infty)$, the test

$$\psi(\tau_0, \tau_1, \tau_2) \equiv \begin{cases} 1, & \text{if } \tau_1 \geq c \\ 0, & \text{if } \tau_1 < c, \end{cases} \tag{6.11}$$

with c chosen as the upper α fractional point of $Q_{0|\tau_0,\tau_2}$, is UMP(α) for testing

$$H_{\tau_0,\tau_2} : \eta = 0 \quad \text{versus} \quad A_{\tau_0,\tau_2} : \eta > 0.$$

Show that for ψ as defined by (6.11), the test

$$\psi(P_{M_0}y, \|P_{M_1}y\|, \|y\|^2)$$

is the size α F-test.

[Hint: For the first assertion, use part **C** and the Neyman-Pearson fundamental lemma. For the second assertion, use the independence under P_θ of $P_{M_0}Y$, $P_{M_1}Y$, and $P_{M^\perp}Y$ along with the fact that if A and B are independent random variables each having central χ^2 distributions, then A/B is independent of $A + B$.]   ○

The next part completes the proof of Theorem 6.6.

G. Let ϕ be a test of the form (6.9) satisfying the similarity condition (6.7). Show that

$$\int \psi(\tau_0, \tau_1, \tau_2)\, Q_{0|\tau_0,\tau_2}(d\tau_1) = \alpha$$

for all (actually, Lebesgue almost all) τ_0 and τ_2, and deduce that ϕ has no greater power than the size α F-test at each $\theta^* \in \Theta_A^*$.

[Hint: First note that for any $\theta^* \in \Theta^*$,

$$\beta_\phi^*(\theta^*) = E_{\theta^*}^* \big(b_\eta(T_0, T_2)\big),$$

where

$$b_\eta(\tau_0, \tau_2) = \int \psi(\tau_0, \tau_1, \tau_2)\, Q_{\eta|\tau_0,\tau_2}(d\tau_1).$$

Then use part **F** in conjunction with the completeness of the statistic (T_0, T_2) for the family $\{\, \theta^* : \theta^* \in \Theta_H^* \,\}$ (see Section 5.2).]   ○

The last two parts deal with some corollaries of Wald's theorem.

H. Deduce

6.12 Theorem (Hsu's theorem). *The size α F-test is uniformly most powerful among all similar size α tests of (6.1) whose power depends on the parameter $\theta = \left(\mu_{M_0}, \mu_{M_1}, \sigma^2 \right)$ only through the distance $\| \mu_{M_1} \| / \sigma$ from μ to M_0 in σ units.* $\circ$

I. Show that the size α F-test ϕ of (6.1) is *admissible*, in the sense that there is no test $\tilde{\phi}$ such that

$$\beta_{\tilde{\phi}}(\theta) \leq \beta_{\phi}(\theta) \text{ for all } \theta \in \Theta_H \quad \text{and} \quad \beta_{\tilde{\phi}}(\theta) \geq \beta_{\phi}(\theta) \text{ for all } \theta \in \Theta_A,$$

with strict inequality holding for at least one $\theta \in \Theta$. $\circ$

CHAPTER 7

ANALYSIS OF COVARIANCE

In this chapter we formulate the analysis of covariance problem in a coordinate-free setting and work out procedures for point and interval estimation and F-testing of hypotheses. We begin by developing various properties of nonorthogonal projections that are needed for the abstract formulation and subsequent analysis.

1. Preliminaries on nonorthogonal projections

Throughout this section, we suppose

$$K = I + J$$

is a direct sum decomposition of a given inner product space K into disjoint, but not necessarily orthogonal, subspaces I and J. Each $x \in K$ can be represented uniquely in the form

$$x = x_I + x_J \tag{1.1}$$

with $x_I \in I$ and $x_J \in J$. The map

$$A \equiv P_{I;J} \colon x \to x_I \text{ is called } \textit{projection onto } I \textit{ along } J, \text{ and}$$

$$B \equiv P_{J;I} \colon x \to x_J \text{ is called } \textit{projection onto } J \textit{ along } I.$$

In visualizing the action of, say, $P_{I;J}$ on $x \in K$, think of x as being translated parallel to J until reaching I, as illustrated below:

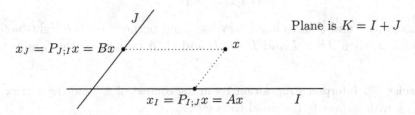

If $I \perp J$, then $P_{I;J} = P_I$ and $P_{J;I} = Q_I$, with P_I and Q_I denoting as usual orthogonal projection onto I and $I^{\perp} = J$, respectively.

1A. Characterization of projections

$P_{I;J}$ may be characterized as follows:

1.2 Proposition. (i) $P_{I;J}$ is an idempotent linear transformation from K to K with range $\mathcal{R}(P_{I;J}) = I$ and null space $\mathcal{N}(P_{I;J}) = J$.

(ii) Conversely, if $T\colon K \to K$ is an idempotent linear transformation with range $\mathcal{R}(T)$ and null space $\mathcal{N}(T)$, then K is the direct sum of $\mathcal{R}(T)$ and $\mathcal{N}(T)$ and $T = P_{\mathcal{R}(T);\mathcal{N}(T)}$.

Proof. (i) is immediate. For (ii), write the general $x \in K$ as

$$x = Tx + (x - Tx).$$

$Tx \in \mathcal{R}(T)$ by definition, and $x - Tx \in \mathcal{N}(T)$ because $T(x - Tx) = Tx - T^2x = Tx - Tx = 0$. It follows that $T = P_{\mathcal{R}(T);\mathcal{N}(T)}$, provided K is in fact the direct sum of $\mathcal{R}(T)$ and $\mathcal{N}(T)$. Now the preceding argument shows that K is indeed the sum of $\mathcal{R}(T)$ and $\mathcal{N}(T)$; the sum is direct because (see Lemma 2.2.39) $\mathcal{R}(T)$ and $\mathcal{N}(T)$ are disjoint — if $x = Ty \in \mathcal{N}(T)$, then $x = Ty = T^2y = Tx = 0$. ∎

1.3 Exercise [2]. Deduce the characterization of orthogonal projections given in Proposition 2.2.4 from Proposition 1.2 above. ◇

1B. The adjoint of a projection

Let T denote projection onto I along J. We inquire as to what the adjoint T' of T is. T' is linear and idempotent ($T'T' = (TT)' = T'$), so T' must be projection onto $\mathcal{R}(T')$ along $\mathcal{N}(T')$. By Exercise 2.5.9,

$$\mathcal{N}(T') = \mathcal{R}^{\perp}(T).$$

Since this identity is actually valid for any linear transformation $T\colon K \to K$, we may replace T by T' to get

$$\mathcal{R}(T') = \mathcal{N}^{\perp}(T).$$

This proves

1.4 Proposition. One has

$$(P_{I;J})' = P_{J^{\perp};I^{\perp}}. \tag{1.5}$$

Notice that I and J have been "reversed and perped" on the right-hand side of (1.5). If $I \perp J$, then $J^{\perp} = I$ and $I^{\perp} = J$, and (1.5) reduces to the familiar formula $P_I' = P_I$.

1.6 Exercise [2]. Interpret Proposition 1.4 in the context of $\mathbb{R}^n$ and the matrix X representing $P_{I;J}$ with respect to the usual basis $e^{(1)}, \ldots, e^{(n)}$.
[Hint: Think about what (1.5) asserts about the row and column spaces of X and its transpose.] ◇

1C. An isomorphism between J and $I^\perp$

The next result is illustrated by this picture:

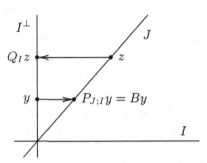

1.7 Proposition. *Let B denote $P_{J;I}$. Then Q_I restricted to J and B restricted to $I^\perp$ are inverses; in particular, J and $I^\perp$ are isomorphic.*

Proof. We need to show

$$\text{(i)}\quad BQ_Iz = z \quad \text{for all } z \in J,$$
$$\text{(ii)}\quad Q_IBy = y \quad \text{for all } y \in I^\perp.$$

For (i), let $z \in J$. Then we have

$$Q_Iz = -P_Iz + z,$$

with $-P_Iz \in I$ and $z \in J$, so $BQ_Iz = z$ as claimed. For (ii), let $y \in I^\perp$. Letting A denote $P_{I;J}$, we have

$$Q_IBy = Q_I(y - Ay) = Q_Iy - Q_IAy = y - 0 = y,$$

as claimed.   ∎

1D. A formula for $P_{J;I}$ when J is given by a basis

If $y_1, \ldots, y_h$ is a basis for $I^\perp$, then we know that, for $x \in K$,

$$Q_Ix = \sum_{1 \le j \le h} c_j y_j$$

can be obtained by solving for the c_j's in the normal equations

$$\left\langle y_i, x - \sum_{1 \le j \le h} c_j y_j \right\rangle = 0 \quad \text{for } i = 1, \ldots, h, \tag{1.8_1}$$

or

$$\langle y_i, x \rangle = \sum_{1 \le j \le h} \langle y_i, y_j \rangle c_j \quad \text{for } i = 1, \ldots, h, \tag{1.8_2}$$

or

$$\langle y, x \rangle = \langle y, y \rangle c, \tag{1.8_3}$$

where the $h \times h$ matrix $\langle y, y \rangle$ and the $h \times 1$ column vectors $\langle y, x \rangle$ and $\mathbf{c}$ are given by

$$(\langle y, x \rangle)_i = \langle y_i, x \rangle, \quad (\langle y, y \rangle)_{ij} = \langle y_i, y_j \rangle, \quad \text{and} \quad (\mathbf{c})_j = c_j \qquad (1.9)$$

for $i, j = 1, \ldots, h$. The solution to (1.8) is of course

$$\mathbf{c} = \langle y, y \rangle^{-1} \langle y, x \rangle.$$

We have just considered the orthogonal case, where $J = I^\perp$. We ask if there is a similar recipe for $P_{J;I} = B$ for general J (disjoint from I). Let $z_1, \ldots, z_h$ be a basis for J. By Proposition 1.7, $Q_I z_1, \ldots, Q_I z_h$ is a basis for $I^\perp$. Now note that Bx depends on $x \in K$ only through $Q_I x$:

$$Bx = B(Q_I x + P_I x) = BQ_I x.$$

We have just seen that

$$Q_I x = \sum_{1 \le j \le h} c_j\, Q_I z_j,$$

where

$$\mathbf{c} = \langle Q_I z, Q_I z \rangle^{-1} \langle Q_I z, x \rangle.$$

And from Proposition 1.7,

$$Bx = BQ_I x = \sum_{1 \le j \le h} c_j\, BQ_I z_j = \sum_{1 \le j \le h} c_j z_j.$$

This gives

1.10 Proposition. Let $z_1, \ldots, z_h$ be a basis for J. For each $x \in K$,

$$P_{J;I} x \equiv Bx = \sum_{1 \le j \le h} c_j z_j, \qquad (1.11)$$

where $c_1, \ldots, c_h$ are the coefficients for the regression of x on $Q_I z_1, \ldots, Q_I z_h$, that is,

$$\mathbf{c} = \langle Q_I z, Q_I z \rangle^{-1} \langle Q_I z, x \rangle \qquad (1.12)$$

(see (1.9) for the notation).

Of course, to use the proposition one needs to know about Q_I in order to compute the numbers

$$\langle Q_I z_i, x \rangle, \quad \langle Q_I z_i, Q_I z_j \rangle, \quad i, j = 1, \ldots, h,$$

appearing in (1.12).

1.13 Exercise [2]. For each $u \in K$, set

$$R(u) = \|Q_I u\|^2 = \|u - P_I u\|^2. \tag{1.14}$$

Verify that for $u, v \in K$,

$$\langle Q_I u, v \rangle = \langle Q_I u, Q_I v \rangle = \frac{R(u+v) - R(u) - R(v)}{2}; \tag{1.15}$$

in particular, if

$$R(u) = \sum_{1 \le j \le m} b_j (\langle u, w_j \rangle)^2 \tag{1.16}$$

for $b_1, \ldots, b_m$ real and $w_1, \ldots, w_m$ in K, then

$$\langle Q_I u, v \rangle = \langle Q_I u, Q_I v \rangle = \sum_{1 \le j \le m} b_j \langle u, w_j \rangle \langle v, w_j \rangle. \tag{1.17} \diamond$$

1E. A formula for $P'_{J;I}$ when J is given by a basis

As above, let $z_1, \ldots, z_h$ be a basis for J and let $Q_I z_1, \ldots, Q_I z_h$ be the corresponding basis for $I^\perp$. Let B denote $P_{J;I}$. By Proposition 1.4, $B' = P_{I^\perp; J^\perp}$. Thus, for $x \in K$,

$$B'x = \sum_{1 \le j \le h} d_j Q_I z_j,$$

where the d_j's are determined by the condition that $x - B'x$ lies in $J^\perp$, that is, by the equations

$$\left\langle z_i, x - \sum_{1 \le j \le h} d_j Q_I z_j \right\rangle = 0 \quad \text{for } i = 1, \ldots, h.$$

Thus

1.18 Proposition. *Let $z_1, \ldots, z_h$ be a basis for J. Then, for each $x \in K$,*

$$P'_{J;I} x \equiv B'x = P_{I^\perp; J^\perp} x = \sum_{1 \le j \le h} d_j Q_I z_j, \tag{1.19}$$

where

$$\boldsymbol{d} = \langle Q_I z, Q_I z \rangle^{-1} \langle z, x \rangle \tag{1.20}$$

(see (1.9) for the notation).

In comparing Propositions 1.10 and 1.18, notice that $P_{J;I} x$ is a linear combination of the z_j's with coefficients involving the inner products of x with the $Q_I z_j$'s, whereas $P'_{J;I} x$ is a linear combination of the $Q_I z_j$'s with coefficients involving the inner products of x with the z_j's.

1.21 Exercise [1]. Verify that the recipes for $P_{J;I} x$ in Proposition 1.10 and for $P'_{J;I} x$ in Proposition 1.18 yield the same result when $I \perp J$. $\qquad \diamond$

2. The analysis of covariance framework

We begin with an example. Consider a GLM consisting of a one-way analysis of variance design coupled with regression on one covariate t:

$$E(Y_{ij}) = \beta_i + \gamma t_{ij} \quad \text{for } j = 1, \ldots, n_i, \ i = 1, \ldots, p; \tag{2.1}$$

here the Y_{ij}'s are uncorrelated random variables of equal variance σ^2, the t_{ij}'s are known fixed numbers, and $\beta_1, \ldots, \beta_p, \gamma$, and σ^2 are unknown parameters (for $p = 1$ the model is just simple linear regression). Notice that

$$\beta_i = E(Y_{ij}) - \gamma t_{ij}$$

measures the expected "yield" of the jth replication of the ith factor, adjusted for the effect of the covariate. The space $\tilde{M}$ of possible means for $\mathbf{Y} = (Y_{ij})_{1 \le j \le n_i, 1 \le i \le p}$ in this problem may be written as

$$\tilde{M} = M + N,$$

where

$$M = \Big\{ \sum\nolimits_{1 \le i \le p} \beta_i \mathbf{v}_i : \beta_1, \ldots, \beta_p \in \mathbb{R} \Big\}$$

is the usual regression space in the one-way ANOVA model $\big((\mathbf{v}_i)_{i'j'} \equiv \delta_{ii'} \big)$ and

$$N = \big[(t_{ij}) \big] \equiv [\mathbf{t}];$$

it is assumed that the covariate is not constant within each of the p groups, that is, that $\mathbf{t} \notin M$, so $\tilde{M}$ is the direct sum of M and N. Of special interest are the β_i's and γ, or, in vector terms,

$$\mu_M = \sum\nolimits_{1 \le i \le p} \beta_i \mathbf{v}_i = P_{M;N}\mu \quad \text{and} \quad \mu_N = \gamma \mathbf{t} = P_{N;M}\mu,$$

where $\mu = E(\mathbf{Y}) = \mu_M + \mu_N$, as illustrated below:

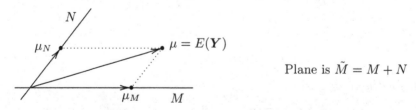

Plane is $\tilde{M} = M + N$

It is natural to seek procedures for making inferences about μ_M and μ_N that build as much as possible on the prior analysis of the one-way ANOVA model.

From a broader perspective, suppose Y is a weakly spherical random variable taking values in an inner product space V:

$$\Sigma_Y = \sigma^2 I_V. \tag{2.2}$$

Suppose also that it is known how to treat the GLM that obtains when $\mu = E(Y)$ is assumed to lie in a certain subspace M of V. The problem at hand is how to use this knowledge advantageously in treating the GLM that arises when the space of possible means of Y is enlarged from M to

$$\tilde{M} = M + N, \tag{2.3}$$

where N is a subspace disjoint from M. For each $\mu \in \tilde{M}$, write

$$\mu = \mu_M + \mu_N \equiv P_{M;N}\mu + P_{N;M}\mu \equiv A\mu + B\mu, \tag{2.4}$$

with $A = P_{M;N}$ and $B = P_{N;M}$ viewed as mappings of $\tilde{M}$ into itself. Generalizing from the specific case above, we presume μ_M and μ_N to be of particular interest.

3. Gauss-Markov estimation

The Gauss-Markov estimator of $\mu \in \tilde{M}$ is

$$\hat{\mu} = P_{\tilde{M}}Y = P_M Y + P_{\tilde{M}-M}Y.$$

The derived Gauss-Markov estimators (see Exercise 4.3.3) $\hat{\mu}_N = B\hat{\mu}$ and $\hat{\mu}_M = A\hat{\mu}$ of $\mu_N = B\mu$ and $\mu_M = A\mu$ may be written as

$$\hat{\mu}_N = BP_{\tilde{M}}Y = BQ_M P_{\tilde{M}}Y = B(P_{\tilde{M}-M}Y) \tag{3.1}$$

and

$$\hat{\mu}_M = P_M\hat{\mu} - P_M\hat{\mu}_N = P_M Y - P_M\hat{\mu}_N, \tag{3.2}$$

as illustrated below:

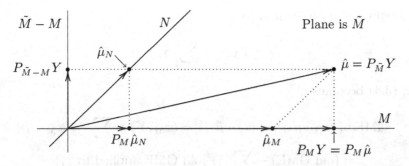

Notice that $\hat{\mu}_M$ is the "old GME" $P_M Y$ (that is, the GME of μ_M in the original problem involving just M) minus the "correction term" $P_M\hat{\mu}_N$, which can be viewed as the old GME applied to $\hat{\mu}_N$ in place of Y. Notice also that $P_M Y$ and $\hat{\mu}_N$ are uncorrelated.

Turning to linear functionals, note that for any $x_N \in N$, the GME of $\langle x_N, \mu_N \rangle = \langle x_N, B\mu \rangle = \langle B'x_N, \mu \rangle$ may be written variously as

$$\langle x_N, \hat{\mu}_N \rangle = \langle x_N, BP_{\tilde{M}}Y \rangle = \langle B'x_N, Y \rangle. \tag{3.3}$$

Similarly, for any $x_M \in M$, the GME of

$$\langle x_M, \mu_M \rangle = \langle x_M, A\mu \rangle = \langle A'x_M, \mu \rangle = \langle x_M - B'x_M, \mu \rangle$$

may be written variously as

$$
\begin{aligned}
\langle x_M, \hat{\mu}_M \rangle &= \langle x_M, P_M Y \rangle - \langle x_M, P_M \hat{\mu}_N \rangle \\
&= \langle x_M, Y \rangle - \langle x_M, \hat{\mu}_N \rangle \\
&= \langle x_M - B'x_M, Y \rangle.
\end{aligned}
\tag{3.4}
$$

The second line of this formula expresses the new GME $\langle x_M, \hat{\mu}_M \rangle$ of $\langle x_M, \mu_M \rangle$ as the old GME $\langle x_M, Y \rangle$ less a subtractive correction that is the old GME applied to $\hat{\mu}_N$ in place of Y.

Once $\hat{\mu}_N = B P_{\tilde{M}-M} Y$ is in hand, the computations reduce to those of the original model, which we presume is well understood. The central problem then is one of getting a handle on $\hat{\mu}_N$. According to the formula $\hat{\mu}_N = B P_{\tilde{M}-M} Y$, what one needs to do is to render N orthogonal to M, project Y onto the resulting space, and transfer the projection back to N. This is readily accomplished in explicit form if we have a basis $z_1, \ldots, z_h$ for N. Indeed, from Proposition 1.10, it follows that

$$\hat{\mu}_N = \sum_{1 \le j \le h} \hat{\gamma}_j z_j, \tag{3.5}$$

where the $\hat{\gamma}_j$'s are the coordinates of $P_{\tilde{M}-M} Y$ with respect to the basis $Q_M z_1, \ldots, Q_M z_h$ for $\tilde{M} - M$:

$$\hat{\gamma} = \langle Q_M z, Q_M z \rangle^{-1} \langle Q_M z, P_{\tilde{M}-M} Y \rangle = \langle Q_M z, Q_M z \rangle^{-1} \langle Q_M z, Y \rangle. \tag{3.6}$$

With this representation for $\hat{\mu}_N$, (3.2) becomes

$$\hat{\mu}_M = P_M Y - \sum_j \hat{\gamma}_j P_M z_j. \tag{3.7}$$

Similarly, (3.4) becomes

$$
\begin{aligned}
\text{GME of } \langle x_M, \mu_M \rangle = \langle x_M, \hat{\mu}_M \rangle &= \langle x_M, Y \rangle - \sum_j \hat{\gamma}_j \langle x_M, z_j \rangle \\
&= (\text{old GME}) - \sum_j \hat{\gamma}_j \, (\text{old GME applied to } z_j)
\end{aligned}
\tag{3.8}
$$

and (3.3) becomes

$$\text{GME of } \langle x_N, \mu_N \rangle = \langle x_N, \hat{\mu}_N \rangle = \sum_j \hat{\gamma}_j \langle x_N, z_j \rangle. \tag{3.9}$$

If we write

$$\mu_N = \sum_j \gamma_j z_j, \tag{3.10}$$

so that γ_j is the coordinate of μ_N with respect to z_j in the $z_1, \dots, z_h$ basis, and if we let $x_N^{(j)} \in N$ be the coefficient vector for the jth-coordinate functional for this basis, so that $\langle x_N^{(j)}, z_i \rangle = \delta_{ij}$ for $i = 1, \dots, h$, then (3.9) says that $\hat{\gamma}_j = \sum_i \hat{\gamma}_i \langle x_N^{(j)}, z_i \rangle$ is in fact the GME of $\gamma_j = \langle x_N^{(j)}, \mu_N \rangle$.

3.11 Exercise [3]. Show how formulas (3.5)–(3.7), with the $\hat{\gamma}_j$'s interpreted as the Gauss-Markov estimators of the γ_j's of (3.10), follow directly via basic properties of Gauss-Markov estimation coupled with orthogonalization of the z_j's with respect to M.
[Hint: Corresponding to the unique representation

$$\mu = \mu_M + \Big(\sum_j \gamma_j z_j \Big)$$

of $\mu \in \tilde{M}$ as the sum of a vector in M and a vector in N, there is the unique representation

$$\mu = \Big(\mu_M + \sum_j \gamma_j P_M z_j \Big) + \Big(\sum_j \gamma_j Q_M z_j \Big)$$

of μ as the sum of a vector in M and a vector in $\tilde{M} - M$.] $\diamond$

3.12 Example. In the model (2.1), where $E(Y_{ij}) = \beta_i + \gamma t_{ij}$, one has for any $\boldsymbol{y} = (y_{ij})$

$$(P_M \boldsymbol{y})_{ij} = \bar{y}_i = \frac{\sum_j y_{ij}}{n_i} \quad \text{and} \quad (Q_M \boldsymbol{y})_{ij} = y_{ij} - \bar{y}_i \,.$$

By (3.6)

$$\hat{\gamma} = \frac{\langle Q_M t, \boldsymbol{Y} \rangle}{\langle Q_M t, Q_M t \rangle} = \frac{\sum_{ij}(t_{ij} - \bar{t}_i) Y_{ij}}{\sum_{ij}(t_{ij} - \bar{t}_i)^2} \tag{3.13$_1$}$$

$$= \frac{\sum_{ij}(t_{ij} - \bar{t}_i)(Y_{ij} - \bar{Y}_i)}{\sum_{ij}(t_{ij} - \bar{t}_i)^2} \tag{3.13$_2$}$$

$$= \sum_i \Big[\frac{\sum_j (t_{ij} - \bar{t}_i)^2}{\sum_{i'j'}(t_{i'j'} - \bar{t}_{i'})^2} \Big] \Big[\frac{\sum_j (t_{ij} - \bar{t}_i) Y_{ij}}{\sum_j (t_{ij} - \bar{t}_i)^2} \Big] \,. \tag{3.13$_3$}$$

Notice how the numerator in (3.13$_2$) can be obtained from the denominator by the technique of Exercise 1.13 and how (3.13$_3$) expresses $\hat{\gamma}$ as a weighted average of the slope estimators one obtains by carrying out simple linear regression within each group, with the weights being proportional to reciprocal variances. By (3.8)

$$\hat{\beta}_i = \Big\langle \frac{\boldsymbol{v}_i}{\|\boldsymbol{v}_i\|^2}, \boldsymbol{Y} \Big\rangle - \hat{\gamma} \Big\langle \frac{\boldsymbol{v}_i}{\|\boldsymbol{v}_i\|^2}, t \Big\rangle = \bar{Y}_i - \hat{\gamma}\bar{t}_i \,. \tag{3.14}$$

In vector terms

$$\hat{\mu}_N = \hat{\gamma} t \quad \text{and} \quad \hat{\mu}_M = \sum_i \hat{\beta}_i \boldsymbol{v}_i \,. \qquad \bullet$$

4. Variances and covariances of GMEs

Under the assumption $\mu = E(Y) \in \tilde{M}$, for any vector $x \in \tilde{M}$ the GME of $\langle x, \mu \rangle$ is $\langle x, Y \rangle$, and for $x_1, x_2 \in \tilde{M}$ one has

$$\text{Cov}\big(\langle x_1, Y \rangle, \langle x_2, Y \rangle\big) = \sigma^2 \langle x_1, x_2 \rangle$$
$$\text{Var}\big(\langle x_i, Y \rangle\big) = \sigma^2 \|x_i\|^2.$$

In (3.3) and (3.4) we saw that

$$\langle x_N, \hat{\mu}_N \rangle = \langle B' x_N, Y \rangle \text{ is the GME of } \langle x_N, \mu_N \rangle = \langle B' x_N, \mu \rangle,$$
$$\langle x_M, \hat{\mu}_M \rangle = \langle x_M - B' x_M, Y \rangle \text{ is the GME of } \langle x_M, \mu_M \rangle = \langle x_M - B' x_M, \mu \rangle$$

for $x_N \in N$ and $x_M \in M$. Consequently,

$$\text{Cov}\big(\langle x_{N,1}, \hat{\mu}_N \rangle, \langle x_{N,2}, \hat{\mu}_N \rangle\big) = \sigma^2 \langle B' x_{N,1}, B' x_{N,2} \rangle \qquad (4.1_1)$$
$$\text{Var}\big(\langle x_N, \hat{\mu}_N \rangle\big) = \sigma^2 \|B' x_N\|^2 \qquad (4.1_2)$$

for $x_{N,1}, x_{N,2}, x_N \in N$, and since $M \perp \mathcal{R}(B')$,

$$\text{Cov}\big(\langle x_{M,1}, \hat{\mu}_M \rangle, \langle x_{M,2}, \hat{\mu}_M \rangle\big) = \sigma^2 \langle x_{M,1} - B' x_{M,1}, x_{M,2} - B' x_{M,2} \rangle$$
$$= \sigma^2 \big(\langle x_{M,1}, x_{M,2} \rangle + \langle B' x_{M,1}, B' x_{M,2} \rangle\big)$$
$$= \sigma^2 (\text{old expression} + \text{correction}) \qquad (4.2_1)$$

$$\text{Var}\big(\langle x_M, \hat{\mu}_M \rangle\big) = \sigma^2 \big(\|x_M\|^2 + \|B' x_M\|^2\big)$$
$$= \sigma^2 (\text{old expression} + \text{correction}) \qquad (4.2_2)$$

for $x_{M,1}, x_{M,2}, x_M \in M$. Formula (4.2_2) reflects the fact that $\langle x_M, \hat{\mu} \rangle$ is the difference between the two uncorrelated random variables $\langle x_M, P_M Y \rangle = \langle x_M, Y \rangle$ and $\langle x_M, \hat{\mu}_N \rangle = \langle B' x_M, Y \rangle$. A similar remark applies to (4.2_1).

Supposing $z_1, \ldots, z_h$ is a basis for N, formulas for the inner products

$$\langle B' x_1, B' x_2 \rangle$$

figuring in (4.1) and (4.2) can be obtained easily with the aid of Proposition 1.18. According to that result, for any $x \in \tilde{M}$ one has

$$B' x = \sum_{1 \leq j \leq h} d_j Q_M z_j, \qquad (4.3_1)$$

where

$$d = C \langle z, x \rangle, \qquad (4.3_2)$$

with

$$C = R^{-1} \quad \text{for} \quad R = \langle Q_M z, Q_M z \rangle. \qquad (4.3_3)$$

For $x_1, x_2 \in \tilde{M}$, after setting

$$\boldsymbol{d}^{(i)} = \boldsymbol{C}\langle z, x_i \rangle \quad \text{for } i = 1, 2$$

we thus have

$$
\begin{aligned}
\langle B'x_1, B'x_2 \rangle &= \Big\langle \sum_j d_j^{(1)} Q_M z_j, \sum_k d_k^{(2)} Q_M z_k \Big\rangle \\
&= \sum_{j,k} d_j^{(1)} \langle Q_M z_j, Q_M z_k \rangle d_k^{(2)} \\
&= (\boldsymbol{d}^{(1)})^T \boldsymbol{R} \boldsymbol{d}^{(2)} = \langle z, x_1 \rangle^T \boldsymbol{C} \langle z, x_2 \rangle.
\end{aligned}
\tag{4.4}
$$

In particular,

$$\|B'x\|^2 = \boldsymbol{d}^T \boldsymbol{R} \boldsymbol{d} = \langle z, x \rangle^T \boldsymbol{C} \langle z, x \rangle \quad \text{with } \boldsymbol{d} = \boldsymbol{C} \langle z, x \rangle \tag{4.5}$$

for $x \in \tilde{M}$.

To see the probabilistic significance of these algebraic manipulations, write as before

$$\hat{\mu}_N = \sum_j \hat{\gamma}_j z_j,$$

and again let $x_N^{(j)}$ be the coefficient vector of the jth-coordinate functional for the basis $z_1, \ldots, z_h$, so $\hat{\gamma}_j = \langle x_N^{(j)}, \hat{\mu}_N \rangle$. Then (4.1_1) and (4.4) give

$$
\begin{aligned}
\text{Cov}(\hat{\gamma}_j, \hat{\gamma}_k) &= \sigma^2 \langle B' x_N^{(j)}, B' x_N^{(k)} \rangle \\
&= \sigma^2 \langle z, x_N^{(j)} \rangle^T \boldsymbol{C} \langle z, x_N^{(k)} \rangle \\
&= \sigma^2 C_{jk};
\end{aligned}
$$

in other words,

$$\sigma^2 \boldsymbol{C} \text{ is the variance-covariance matrix of } \hat{\gamma}, \tag{4.6}$$

as could have been anticipated from (3.6) (see (4.2.22) and (4.2.24)). Since, for any $x_1, x_2 \in \tilde{M}$,

$$\langle x_i, \hat{\mu}_N \rangle = \sum_j \langle z_j, x_i \rangle \hat{\gamma}_j,$$

(4.4) and the rule for figuring the covariance of two linear combinations of a given set of random variables give

$$
\begin{aligned}
\sigma^2 \langle B'x_1, B'x_2 \rangle &= \sigma^2 \langle z, x_1 \rangle^T \boldsymbol{C} \langle z, x_2 \rangle \\
&= \text{Cov}\Big(\sum_j \langle z_j, x_1 \rangle \hat{\gamma}_j, \sum_j \langle z_j, x_2 \rangle \hat{\gamma}_j \Big) \\
&= \text{Cov}(\langle x_1, \hat{\mu}_N \rangle, \langle x_2, \hat{\mu}_N \rangle).
\end{aligned}
\tag{4.7}
$$

4.8 Exercise [2]. What is the covariance of $\langle x_M, \hat{\mu}_M \rangle$ and $\langle x_N, \hat{\mu}_N \rangle$ for given $x_M \in M$ and $x_N \in N$? ◇

4.9 Example. Consider again our running example, where $E(Y_{ij}) = \beta_i + \gamma t_{ij}$. From (4.6) and (4.3$_3$),

$$\text{Var}(\hat{\gamma}) = \sigma^2 C_{11} = \frac{\sigma^2}{\sum_{ij}(t_{ij} - \bar{t}_i)^2}, \tag{4.10}$$

as one would obtain directly from

$$\hat{\gamma} = \frac{\sum_{ij}(t_{ij} - \bar{t}_i)Y_{ij}}{\sum_{ij}(t_{ij} - \bar{t}_i)^2}.$$

Moreover, from (4.2$_2$) and (4.5), for any $c_1, \ldots, c_p \in \mathbb{R}$ we have, upon writing $\boldsymbol{x}$ for $\sum_i c_i \boldsymbol{v}_i / \|\boldsymbol{v}_i\|^2$,

$$
\begin{aligned}
\text{Var}\Big(\text{GME of} \sum_i c_i \beta_i\Big) &= \text{Var}(\langle \boldsymbol{x}, \hat{\mu}_M \rangle) \\
&= \sigma^2 \big(\|\boldsymbol{x}\|^2 + \|B'\boldsymbol{x}\|^2\big) \\
&= \sigma^2 \Big(\sum_i \frac{c_i^2}{n_i} + \langle \boldsymbol{t}, \boldsymbol{x} \rangle^T C \langle \boldsymbol{t}, \boldsymbol{x} \rangle\Big) \\
&= \sigma^2 \Big(\sum_i \frac{c_i^2}{n_i} + \frac{\big(\sum_i c_i \bar{t}_i\big)^2}{\sum_{ij}(t_{ij} - \bar{t}_i)^2}\Big),
\end{aligned}
\tag{4.11}
$$

as one would obtain directly from

$$\text{GME of} \sum_i c_i \beta_i = \text{old GME} - \text{old GME applied to } \hat{\mu}_N$$

$$= \Big(\sum_i c_i \bar{Y}_i\Big) - \hat{\gamma} \sum_i c_i \bar{t}_i$$

(recall that $P_M Y$ and $\hat{\mu}_N$ are uncorrelated). •

4.12 Exercise [1]. Show that when $p = 1$, formula (4.10) reduces to the usual expression for the variance of the GME of the slope coefficient in simple linear regression, and (4.11) with $c_1 = 1$ reduces to the usual expression for the variance of the GME of the intercept.◇

5. Estimation of σ^2

The standard unbiased estimator of σ^2 is

$$\hat{\sigma}^2 = \frac{\|Q_{\tilde{M}} Y\|^2}{d(\tilde{M}^\perp)}.$$

We have

$$d(\tilde{M}) = d(M) + d(N)$$

because M and N are disjoint, so

$$d(\tilde{M}^\perp) = d(M^\perp) - d(N)$$

$$= \text{old degrees of freedom} - \text{correction term}. \tag{5.1}$$

The Pythagorean theorem applied to the orthogonal decomposition

$$P_{\tilde{M}-M}Y + Q_{\tilde{M}}Y = Q_M Y$$

gives

$$\|Q_{\tilde{M}}Y\|^2 = \|Q_M Y\|^2 - \|P_{\tilde{M}-M}Y\|^2$$

$$= \text{old term} - \text{correction term}.$$

(5.2)

If $z_1, \ldots, z_h$ is a basis for N, so that $Q_M z_1, \ldots, Q_M z_h$ is a basis for $\tilde{M} - M$, then

$$P_{\tilde{M}-M}Y = \sum_j \hat{\gamma}_j Q_M z_j,$$

as pointed out above (3.6), and

$$\|P_{\tilde{M}-M}Y\|^2 = \sum_{jk} \hat{\gamma}_j \langle Q_M z_j, Q_M z_k \rangle \hat{\gamma}_k$$

$$= \hat{\gamma}^T \boldsymbol{R} \hat{\gamma}$$

$$= \langle Q_M z, Y \rangle^T \boldsymbol{C} \langle Q_M z, Y \rangle,$$

(5.3)

with the matrices $\boldsymbol{R}$ and $\boldsymbol{C}$ being defined by (4.3$_3$).

5.4 Example. In our running example with $E(Y_{ij}) = \beta_i + \gamma t_{ij}$, set

$$n = \sum_{1 \leq i \leq p} n_i.$$

One has

$$d(\tilde{M}^\perp) = d(M^\perp) - d(N) = (n - p) - 1$$

and

$$\|Q_{\tilde{M}}Y\|^2 = \|Q_M Y\|^2 - \|P_{\tilde{M}-M}Y\|^2 = \sum_{ij}(Y_{ij} - \bar{Y}_i)^2 - \hat{\gamma}^2 \sum_{ij}(t_{ij} - \bar{t}_i)^2. \quad \bullet$$

5.5 Exercise [1]. Show that an alternate expression for $\|Q_{\tilde{M}}Y\|^2$ in the running example is

$$\sum_{ij}(Y_{ij} - \bar{Y}_i - \hat{\gamma}(t_{ij} - \bar{t}_i))^2. \qquad \diamond$$

6. Scheffé intervals for functionals of μ_M

Throughout this section suppose that Y is normally distributed. For any vector space, say $\tilde{\mathcal{L}}$, of linear functionals $\tilde{\psi}$ of $\mu \in \tilde{M}$, Scheffé's result says

$$P_{\mu,\sigma^2}\left(\tilde{\psi}(P_{\tilde{M}}Y) - S\hat{\sigma}_{\tilde{\psi}(P_{\tilde{M}}Y)} \leq \tilde{\psi}(\mu) \leq \tilde{\psi}(P_{\tilde{M}}Y) + S\hat{\sigma}_{\tilde{\psi}(P_{\tilde{M}}Y)}\right.$$

$$\left. \text{for all } \tilde{\psi} \in \tilde{\mathcal{L}}\right) = 1 - \alpha$$

(6.1)

for all $\mu \in \tilde{M}$ and $\sigma^2 > 0$; here

$$\hat{\sigma}_{\tilde{\psi}(P_{\tilde{M}}Y)} = \hat{\sigma}\| \, cv\,\tilde{\psi}\|, \quad \hat{\sigma}^2 = \frac{\|Q_{\tilde{M}}Y\|^2}{d(\tilde{M}^\perp)}, \quad \text{and} \quad S = S_{d(\tilde{\mathcal{L}}),d(\tilde{M}^\perp)}(\alpha).$$

We want to particularize (6.1) to the case where $\tilde{\mathcal{L}}$ is essentially a space of functionals of μ_M.

Let then $\mathcal{L}$ be a vector space of functionals on M. For each $\psi \in \mathcal{L}$, let $x_\psi \in M$ be the coefficient vector of ψ:

$$\psi(m) = \langle m, x_\psi \rangle \quad \text{for all } m \in M.$$

Each $\psi \in \mathcal{L}$ gives rise to a corresponding linear functional $\tilde{\psi}$ on $\tilde{M}$ via the recipe

$$\tilde{\psi}(\mu) = \psi(\mu_M) = \psi(A\mu) = \langle x_\psi, A\mu \rangle = \langle A'x_\psi, \mu \rangle$$

for $\mu \in \tilde{M}$. Let $\tilde{\mathcal{L}}$ be the space $\{\, \tilde{\psi} : \psi \in \mathcal{L} \,\}$. The spaces of coefficient vectors

$$L = \{\, x_\psi : \psi \in \mathcal{L} \,\} \subset M$$
$$\tilde{L} = \{\, A'x_\psi : \psi \in \mathcal{L} \,\} \subset \tilde{M}$$

are related by

$$\tilde{L} = A'(L).$$

Now A', being projection onto $\tilde{M} - N$ along $\tilde{M} - M$ within $\tilde{M}$, has null space $\tilde{M} - M$ and so is nonsingular on L. Thus

$$d(\tilde{\mathcal{L}}) = d(\tilde{L}) = d(L) = d(\mathcal{L}).$$

It follows from (6.1) that

$$P_{\mu,\sigma^2}\big(\psi(\hat{\mu}_M) - S\hat{\sigma}_{\psi(\hat{\mu}_M)} \leq \psi(\mu_M) \leq \psi(\hat{\mu}_M) + S\hat{\sigma}_{\psi(\hat{\mu}_M)}$$
$$\text{for all } \psi \in \mathcal{L}\big) = 1 - \alpha \tag{6.2}$$

for all $\mu \in \tilde{M}$ and $\sigma^2 > 0$, with

$$S = S_{d(\mathcal{L}),d(\tilde{M}^\perp)}(\alpha), \tag{6.3$_1$}$$

$$\hat{\sigma}_{\psi(\hat{\mu}_M)} = \hat{\sigma}\|A'x_\psi\| = \hat{\sigma}\sqrt{\|x_\psi\|^2 + \|B'x_\psi\|^2}, \tag{6.3$_2$}$$

$$\hat{\sigma}^2 = \frac{\|Q_{\tilde{M}}Y\|^2}{d(\tilde{M}^\perp)}. \tag{6.3$_3$}$$

6.4 Example. In our running example with $E(Y_{ij}) = \beta_i + \gamma t_{ij}$, consider the space $\mathcal{L}$ of contrasts in the β_i's. We know

$$d(\mathcal{L}) = p - 1$$

and we have seen that for any contrast $\psi(\mu_M) = \sum_i c_i \beta_i$,

$$\psi(\hat{\mu}_M) = \sum_i c_i (\bar{Y}_i - \hat{\gamma}\bar{t}_i)$$

$$\hat{\sigma}_{\psi(\hat{\mu}_M)} = \hat{\sigma} \left(\sum_i \frac{c_i^2}{n_i} + \frac{\left(\sum_i c_i \bar{t}_i\right)^2}{\sum_{ij}(t_{ij} - \bar{t}_i)^2} \right)^{1/2}$$

with

$$\hat{\sigma}^2 = \frac{\sum_{ij}(Y_{ij} - \bar{Y}_i)^2 - \hat{\gamma}^2 \sum_{ij}(t_{ij} - \bar{t}_i)^2}{(n - p) - 1}.$$

Thus the intervals

$$\psi(\hat{\mu}_M) \pm S_{p-1,n-p-1}(\alpha)\, \hat{\sigma}_{\psi(\hat{\mu}_M)}$$

have simultaneous confidence $100(1 - \alpha)\%$ for the corresponding contrasts $\psi(\mu_M)$ for $\psi \in \mathcal{L}$.

6.5 Exercise [3]. What are the Scheffé intervals for a vector space $\mathcal{L}$ of functionals of μ_N in the general augmented model ($\mu \in \tilde{M}$)? ◇

6.6 Exercise [3]. An experiment is contemplated in which an even number $n = 2m$ of observations are to be taken at equally spaced time intervals. For $1 \le i \le n$, the response Y_i for the ith observation is assumed to take the form

$$Y_i = \begin{cases} \alpha + \Delta + \epsilon_i, & \text{if } i \text{ is odd,} \\ \alpha - \Delta + \epsilon_i, & \text{if } i \text{ is even,} \end{cases} \tag{6.7}$$

where α is the mean output of the experimental apparatus in the absence of any treatment effect, Δ (respectively, $-\Delta$) is the incremental mean response due to a treatment applied in a positive (respectively, negative) direction, and the ϵ_i's are experimental errors. α is of no interest. (1) Assuming that the ϵ_i's are uncorrelated with zero means and common unknown variance $\sigma^2 > 0$, give explicit formulas for the GME of Δ and its estimated standard deviation; also set $100(1 - \alpha)\%$ confidence limits on Δ. (2) The proposed model (6.7) is defective in that it does not take into account a possible drift of the experimental apparatus over time. As an improved model one might take

$$Y_i = \begin{cases} \alpha + \beta t_i + \Delta + \epsilon_i, & \text{if } i \text{ is odd,} \\ \alpha + \beta t_i - \Delta + \epsilon_i, & \text{if } i \text{ is even,} \end{cases}$$

where $t_i = i t_1$ is the time at which the ith observation is to be taken. What now are the GME of Δ, its estimated standard deviation, and the confidence limits? ◇

7. *F*-testing

To motivate things, consider our running example with $E(Y_{ij}) = \beta_i + \gamma t_{ij}$. The hypothesis

$$H: \beta_1 = \beta_2 = \cdots = \beta_p$$

constrains $\mu \in \tilde{M}$ to lie in

$$\tilde{M}_0 = M_0 + N,$$

where

$$M_0 = \Big\{ \sum_i \beta_i \boldsymbol{v}_i : \beta_1 = \beta_2 = \cdots = \beta_p \Big\} = [\boldsymbol{e}]$$

is the hypothesis space in the original one-way ANOVA model.

In general, if one has a GLM that presupposes $\mu \in M$ and is interested in testing

$$H: \mu \in M_0 \quad \text{versus} \quad A: \mu \notin M_0, \tag{7.1}$$

then in the augmented model, where $\mu \in \tilde{M} = M + N$, the corresponding testing problem is

$$H: \mu \in \tilde{M}_0 \quad \text{versus} \quad A: \mu \notin \tilde{M}_0, \tag{7.2}$$

where

$$\tilde{M}_0 = M_0 + N;$$

the sum here is direct because we have assumed N to be disjoint from M. The F-statistic for (7.1) is

$$F = \frac{\big(\|Q_{M_0} Y\|^2 - \|Q_M Y\|^2 \big) / d(M - M_0)}{\|Q_M Y\|^2 / d(M^\perp)}, \tag{7.3}$$

while that for (7.2) is

$$\tilde{F} = \frac{\big(\|Q_{\tilde{M}_0} Y\|^2 - \|Q_{\tilde{M}} Y\|^2 \big) / d(\tilde{M} - \tilde{M}_0)}{\|Q_{\tilde{M}} Y\|^2 / d(\tilde{M}^\perp)}. \tag{7.4}$$

From our previous work we know how each term in (7.4) can be obtained from the corresponding term in (7.3) via an appropriate correction:

$$\|Q_{\tilde{M}} Y\|^2 = \|Q_M Y\|^2 - \|P_{\tilde{M} - M} Y\|^2, \tag{7.5_1}$$

with

$$\|P_{\tilde{M} - M} Y\|^2 = \hat{\boldsymbol{\gamma}}_M^T \boldsymbol{R}_M \hat{\boldsymbol{\gamma}}_M, \quad \boldsymbol{R}_M = \langle Q_M z, Q_M z \rangle, \quad \hat{\boldsymbol{\gamma}}_M = \boldsymbol{R}_M^{-1} \langle Q_M z, Y \rangle$$

in the basis case;

$$\|Q_{\tilde{M}_0} Y\|^2 = \|Q_{M_0} Y\|^2 - \|P_{\tilde{M}_0 - M_0} Y\|^2, \tag{7.5_2}$$

with

$$\|P_{\tilde{M}_0 - M_0} Y\|^2 = \hat{\boldsymbol{\gamma}}_{M_0}^T \boldsymbol{R}_{M_0} \hat{\boldsymbol{\gamma}}_{M_0}, \quad \boldsymbol{R}_{M_0} = \langle Q_{M_0} z, Q_{M_0} z \rangle,$$
$$\hat{\boldsymbol{\gamma}}_{M_0} = \boldsymbol{R}_{M_0}^{-1} \langle Q_{M_0} z, Y \rangle$$

in the basis case;

$$d(\tilde{M}^\perp) = d(M^\perp) - d(N); \tag{7.5_3}$$

and

$$d(\tilde{M} - \tilde{M}_0) = d(\tilde{M}) - d(\tilde{M}_0)$$
$$= \big(d(M) + d(N) \big) - \big(d(M_0) + d(N) \big)$$

$$= d(M - M_0). \tag{7.5_4}$$

7.6 Example. Consider again the case

$$E(Y_{ij}) = \beta_i + \gamma t_{ij}, \quad j = 1, \ldots, n_i, \quad i = 1, \ldots, p,$$

with

$$H: \beta_1 = \beta_2 = \cdots = \beta_p.$$

Here

$$(Q_M \boldsymbol{Y})_{ij} = Y_{ij} - \bar{Y}_i, \qquad \|Q_M \boldsymbol{Y}\|^2 = \sum_{ij}(Y_{ij} - \bar{Y}_i)^2$$
$$(Q_{M_0} \boldsymbol{Y})_{ij} = Y_{ij} - \bar{Y}, \qquad \|Q_{M_0} \boldsymbol{Y}\|^2 = \sum_{ij}(Y_{ij} - \bar{Y})^2$$

with $\bar{Y} = \sum_{ij} Y_{ij} / \sum_i n_i$. One has

$$\|Q_{M_0} \boldsymbol{Y}\|^2 - \|Q_M \boldsymbol{Y}\|^2 = \|P_{M-M_0} \boldsymbol{Y}\|^2 = \sum_i n_i(\bar{Y}_i - \bar{Y})^2,$$

$$\|P_{\tilde{M}-M} \boldsymbol{Y}\|^2 = \hat{\gamma}_M^2 \|Q_M \boldsymbol{t}\|^2 = \hat{\gamma}_M^2 \sum_{ij}(t_{ij} - \bar{t}_i)^2,$$

with

$$\hat{\gamma}_M = \frac{\langle Q_M \boldsymbol{t}, \boldsymbol{Y} \rangle}{\|Q_M \boldsymbol{t}\|^2} = \frac{\sum_{ij}(t_{ij} - \bar{t}_i)Y_{ij}}{\sum_{ij}(t_{ij} - \bar{t}_i)^2}$$

as in (3.13) and

$$\|P_{\tilde{M}_0-M_0} \boldsymbol{Y}\|^2 = \hat{\gamma}_{M_0}^2 \|Q_{M_0} \boldsymbol{t}\|^2 = \hat{\gamma}_{M_0}^2 \sum_{ij}(t_{ij} - \bar{t})^2,$$

with

$$\hat{\gamma}_{M_0} = \frac{\langle Q_{M_0} \boldsymbol{t}, \boldsymbol{Y} \rangle}{\|Q_{M_0} \boldsymbol{t}\|^2} = \frac{\sum_{ij}(t_{ij} - \bar{t})Y_{ij}}{\sum_{ij}(t_{ij} - \bar{t})^2}$$

as in simple linear regression, whence the *F*-statistic is

$$\frac{\left[\sum_i n_i(\bar{Y}_i - \bar{Y})^2 - \left[\hat{\gamma}_{M_0}^2 \sum_{ij}(t_{ij} - \bar{t})^2 - \hat{\gamma}_M^2 \sum_{ij}(t_{ij} - \bar{t}_i)^2 \right] \right] \Big/ (p-1)}{\left[\sum_{ij}(Y_{ij} - \bar{Y}_i)^2 - \hat{\gamma}_M^2 \sum_{ij}(t_{ij} - \bar{t}_i)^2 \right] \Big/ ((n-p)-1)}.$$

7.7 Exercise [2]. Show that an alternative expression for the numerator of the *F*-statistic in the preceding example is

$$\sum_{ij}(\bar{Y}_i - \bar{Y} + \hat{\gamma}_M(t_{ij} - \bar{t}_i) - \hat{\gamma}_{M_0}(t_{ij} - \bar{t}))^2 / (p-1). \qquad \diamond$$

7.8 Exercise [3]. Let M_0 be a subspace of M. (1) Find the *F*-statistic for testing

$$H: P_M \mu \in M_0 \quad \text{versus} \quad A: P_M \mu \notin M_0 \tag{7.9}$$

in the context of the general augmented model, where $\mu \in \tilde{M} = M + N$. (2) In what respects does the F-statistic in (1) differ from that for the problem (7.2), and from that for the problem (7.1) in the context of the original model, where $\mu \in M$? (3) Verify that in the context of our running example, where $E(Y_{ij}) = \beta_i + \gamma t_{ij}$, the hypothesis H in (7.9) is that the unadjusted mean yields $\beta_i + \gamma \bar{t}_i$, $i = 1, \ldots, p$, are equal.   ◇

7.10 Exercise [2]. In the general augmented problem ($\mu \in \tilde{M}$), what is the F-statistic for the null hypothesis that $\mu_N = 0$?   ◇

Assuming the normality of Y, the power of the F-test of

$$H : \mu \in \tilde{M}_0 \quad \text{versus} \quad A : \mu \notin \tilde{M}_0$$

(presuming $\mu \in \tilde{M}$) at (μ_M, μ_N, σ^2) is

$$\mathcal{F}_{d(M-M_0),\, d(M^{\perp})-d(N);\, \delta}\Big(\big[\mathcal{F}_{d(M-M_0),d(M^{\perp})-d(N)}(\alpha), \infty \big) \Big), \qquad (7.11)$$

where the noncentrality parameter δ is determined by the relation

$$
\begin{aligned}
\sigma^2 \delta^2 &= \| Q_{\tilde{M}_0} \mu \|^2 \\
&= \| Q_{\tilde{M}_0} \mu_M + Q_{\tilde{M}_0} \mu_N \|^2 & (\mu = \mu_M + \mu_N) \\
&= \| Q_{\tilde{M}_0} \mu_M \|^2 & (\mu_N \in N \subset M_0 + N = \tilde{M}_0) \\
&= \| Q_{M_0} \mu_M \|^2 - \| P_{\tilde{M}_0 - M_0} \mu_M \|^2 & (\text{see } (7.5_2)) \\
&= \| Q_{M_0} \mu_M \|^2 - \| P_{\tilde{M}_0 - M_0} Q_{M_0} \mu_M \|^2 & \big(P_{M_0}\mu_M \perp (\tilde{M}_0 - M_0) \big),
\end{aligned}
$$

that is,

$$\delta = \sqrt{ \frac{\| Q_{M_0} \mu_M \|^2}{\sigma^2} - \frac{\| P_{\tilde{M}_0 - M_0} Q_{M_0} \mu_M \|^2}{\sigma^2} }$$

$$= \sqrt{ \text{old expression} - \text{correction term}}. \qquad (7.12)$$

Notice that δ does not depend on μ_N and depends on μ_M only through $Q_{M_0}\mu_M = P_{M-M_0}\mu_M$. In the case that $z_1, \ldots, z_h$ is a basis for N,

$$\| P_{\tilde{M}_0 - M_0} Q_{M_0} \mu_M \|^2 = \langle Q_{M_0} z, Q_{M_0} \mu_M \rangle^T \langle Q_{M_0} z, Q_{M_0} z \rangle^{-1} \langle Q_{M_0} z, Q_{M_0} \mu_M \rangle$$

$$= \frac{\big(\langle Q_{M_0} z_1, Q_{M_0} \mu_M \rangle \big)^2}{\langle Q_{M_0} z_1, Q_{M_0} z_1 \rangle}, \quad \text{if } h = 1.$$

7.13 Example. In connection with testing $H : \beta_1 = \beta_2 = \cdots = \beta_p$ in our running example ($E(Y_{ij}) = \beta_i + \gamma t_{ij}$), one has

$$\mu_M = \sum_i \beta_i v_i, \quad M_0 = [e], \quad Q_{M_0}\mu_M = \sum_i (\beta_i - \bar{\beta}) v_i,$$

with $\bar{\beta}$ being the weighted average $\left(\sum_i n_i \beta_i\right) / \sum_i n_i$, whence

$$\sigma^2 \delta^2 = \sum_i n_i (\beta_i - \bar{\beta})^2 - \frac{\left(\sum_i n_i (\beta_i - \bar{\beta})(\bar{t}_i - \bar{t})\right)^2}{\sum_{ij}(t_{ij} - \bar{t})^2}.$$

$\bullet$

7.14 Exercise [1]. In the context of Exercise 6.6, what is the F-test of the hypothesis $H: \Delta = 0$ versus the alternative $A: \Delta \neq 0$, and what is the power of that test? $\diamond$

8. Problem set: The Latin square design

In this problem set you are asked to work through the analysis of a Latin square design. Clarity of exposition is especially important. Ideas and techniques developed throughout this book are called upon and should be employed wherever possible. Parts **A–H** can be worked using just the material from Chapters 1–6, part **I** requires this chapter, and part **J** requires the next chapter.

Consider an additive three-way ANOVA design with m levels for each of the three classifications and one observation per cell. This design calls for m^3 observations, which may be beyond the resources available for the experiment. An easily analyzed incomplete design for handling this problem is the so-called *Latin square design (LSD)*, which requires only m^2 observations.

An LSD is an incomplete three-way layout in which all three factors have, say, $m \geq 2$ levels and in which one observation is taken on each of m^2 treatment combinations in such a way that each level of each factor is combined exactly once with each level of each of the other two factors. The m^2 treatment combinations ijk are usually presented in an $m \times m$ array, in which the rows correspond to levels of the first factor, say A, the columns to levels of the second factor, say B, and the entries to the levels of the third factor, say C. The rule for combining treatments requires that each integer $k = 1, \ldots, m$ appear exactly once in each row and exactly once in each column. An $m \times m$ array with this property, such as

$$
\begin{array}{cccc}
1 & 2 & 3 & 4 \\
2 & 3 & 4 & 1 \\
3 & 4 & 1 & 2 \\
4 & 1 & 2 & 3
\end{array}
\tag{8.1}
$$

for $m = 4$, is called an $m \times m$ *Latin square*. Each $m \times m$ Latin square can be used to determine an $m \times m$ LSD.

So let D be a given $m \times m$ Latin square. Set

$$\mathcal{D} = \{\, ijk \in \{1, 2, \ldots, m\}^3 : k \text{ is the entry in the } i^{\text{th}} \text{ row}$$
$$\text{and the } j^{\text{th}} \text{ column of } D \,\}.$$

Notice that a triple ijk in $\mathcal{D}$ is determined by any two of its components. For example, with D as in (8.1), the triple in $\mathcal{D}$ with $i = 3$ and $k = 2$ is 342. We will be dealing with the Latin square design determined by D: Specifically, we will be dealing with uncorrelated random variables Y_{ijk}, $ijk \in \mathcal{D}$, that have common (unknown) variance σ^2 and mean structure of the form

$$E(Y_{ijk}) = \nu + \alpha_i^A + \alpha_j^B + \alpha_k^C \tag{8.2}$$

for some *grand mean* ν and *differential treatment effects* α_i^A, α_j^B, α_k^C, $i, j, k = 1, \ldots, m$ satisfying the constraints

$$\alpha_.^A = \alpha_.^B = \alpha_.^C = 0. \tag{8.3}$$

Here $\alpha_.^G$ denotes the arithmetic average of the quantities $\alpha_1^G, \ldots, \alpha_m^G$ for $G = A, B, C$.

Let

$$V = \{ (y_{ijk})_{ijk \in \mathcal{D}} : y_{ijk} \in \mathbb{R} \text{ for each } ijk \in \mathcal{D} \}$$

be endowed with the dot-product, and let M be the regression space determined by (8.2) and (8.3):

$$M = \{ y \in V : y_{ijk} = \nu + \alpha_i^A + \alpha_j^B + \alpha_k^C \text{ for some}$$
$$\nu, \alpha_i^A\text{'s}, \alpha_j^B\text{'s, and } \alpha_k^C\text{'s satisfying } \alpha_.^A = \alpha_.^B = \alpha_.^C = 0 \}.$$

Let e and x_i^A, x_j^B, x_k^C for $i, j, k = 1, \ldots, m$ be the vectors in V defined by

$$(e)_{i'j'k'} = 1, \quad (x_i^A)_{i'j'k'} = \delta_{ii'}, \quad (x_j^B)_{i'j'k'} = \delta_{jj'}, \quad (x_k^C)_{i'j'k'} = \delta_{kk'}$$

for $i'j'k' \in \mathcal{D}$.

A. Show that

$$M = \text{Span}\{ e, x_1^A, \ldots, x_m^A, x_1^B, \ldots, x_m^B, x_1^C, \ldots, x_m^C \}$$

and verify

$$\begin{aligned}
\|e\|^2 &= m^2, \\
\|x_g^G\|^2 &= m, \quad \text{for } G = A, B, C \text{ and } g = 1, 2, \ldots, m \\
\langle x_g^G, e \rangle &= m, \quad \text{for } G = A, B, C \text{ and } g = 1, 2, \ldots, m \\
\langle x_g^G, x_h^G \rangle &= 0, \quad \text{for } G = A, B, C \text{ and } g \neq h, \\
\langle x_g^G, x_h^H \rangle &= 1, \quad \text{for } G \neq H \text{ and } g, h = 1, \ldots, m.
\end{aligned}$$

○

B. For $\mu \in M$, define

$$\mu_{\ldots} = \frac{1}{m^2} \sum_{i'j'k' \in \mathcal{D}} \mu_{i'j'k'}$$

$$\mu_{i\cdot\cdot} = \frac{1}{m} \sum_{\substack{i'j'k' \in \mathcal{D} \\ i'=i}} \mu_{i'j'k'}, \quad \mu_{\cdot j\cdot} = \frac{1}{m} \sum_{\substack{i'j'k' \in \mathcal{D} \\ j'=j}} \mu_{i'j'k'}, \quad \mu_{\cdot\cdot k} = \frac{1}{m} \sum_{\substack{i'j'k' \in \mathcal{D} \\ k'=k}} \mu_{i'j'k'}$$

for $i, j, k = 1, \ldots, m$. These quantities are linear functionals of μ and so may be represented in the form

$$\langle x, \mu \rangle$$

for appropriate coefficient vectors $x \in M$. Express these coefficient vectors in terms of e and the x_i^A's, x_j^B's, and x_k^C's.

Also, show that when μ is written as the tuple

$$(\mu_{ijk})_{ijk \in \mathcal{D}} = (\nu + \alpha_i^A + \alpha_j^B + \alpha_k^C)_{ijk \in \mathcal{D}}$$

with

$$\alpha_.^A = \alpha_.^B = \alpha_.^C = 0,$$

then

$$\nu = \mu_{...}$$

and

$$\alpha_i^A = \mu_{i..} - \mu_{...}, \quad \alpha_j^B = \mu_{.j.} - \mu_{...}, \quad \text{and} \quad \alpha_k^C = \mu_{..k} - \mu_{...}.$$

Express the coefficient vectors for ν, α_i^A, α_j^B, and α_k^C in terms of e and the x_i^A's, x_j^B's, and x_k^C's. ○

C. Show that the GMEs of ν, α_i^A, α_j^B, and α_k^C are

$$\hat{\nu} = \bar{Y}_{...}$$

and

$$\hat{\alpha}_i^A = \bar{Y}_{i..} - \bar{Y}_{...}, \quad \hat{\alpha}_j^B = \bar{Y}_{.j.} - \bar{Y}_{...}, \quad \hat{\alpha}_k^C = \bar{Y}_{..k} - \bar{Y}_{...},$$

where $\bar{Y}_{...}$, and so on, are defined like $\mu_{...}$, and so on, but with μ replaced by Y. Also, verify that

$$P_M Y = \hat{\nu} e + \sum_{1 \leq i \leq m} \hat{\alpha}_i^A x_i^A + \sum_{1 \leq j \leq m} \hat{\alpha}_j^B x_j^B + \sum_{1 \leq k \leq m} \hat{\alpha}_k^C x_k^C. \qquad ○$$

D. Give the variances and covariances of $\hat{\nu}$ and the $\hat{\alpha}_i^A$'s, $\hat{\alpha}_j^B$'s, and $\hat{\alpha}_k^C$'s. (You may find it helpful to work part **E** before this one.) ○

Now introduce the canonical null hypotheses

$$H_A: \text{all } \alpha_i^A = 0, \quad H_B: \text{all } \alpha_j^B = 0, \quad \text{and} \quad H_C: \text{all } \alpha_k^C = 0.$$

These hypotheses restrict $\mu \in M$ to subspaces we denote by M_A, M_B, and M_C, respectively. For notational convenience write

$$L_A = M - M_A, \quad L_B = M - M_B, \quad \text{and} \quad L_C = M - M_C.$$

E. Show that

$$L_A = \text{Span}\{ Q_{[e]} x_i^A : i = 1, \ldots, m \}$$
$$= \text{Span}\{ x_i^A : i = 1, \ldots, m \} - [e]$$

and that

$$P_{L_A} Y = \sum_i \bar{Y}_{i\cdot\cdot} x_i^A - \bar{Y}_{\cdot\cdot\cdot} e = \sum_i \hat{\alpha}_i^A x_i^A .$$

Similar formulas hold vis-à-vis L_B and L_C. Show also that

$$M = [e] + L_A + L_B + L_C$$

is an orthogonal decomposition of M. ○

F. Use the foregoing results to complete, and verify all entries for, the following extended ANOVA table.

Effect	Grand mean	Treatment A	Treatment B	Treatment C	Error	Total
Subspace	——	L_A	——	——	——	V
Dimension (d.f.)	——	$m-1$	——	——	——	——
Sum of squares	SS_{GM}	SS_A	SS_B	SS_C	SS_e	SS_{tot}
closed form	$m^2 \bar{Y}_{\cdot\cdot\cdot}^2$	$m\sum_i (\hat{\alpha}_i^A)^2$	——	——	——	$\sum Y_{ijk}^2$
open form		$m\sum_i \bar{Y}_{i\cdot\cdot}^2 - m^2 \bar{Y}_{\cdot\cdot\cdot}^2$	——	——	——	
Mean Square	MS_{GM} ——	MS_A $SS_A/(m-1)$	MS_B ——	MS_C ——	MS_e ——	MS_{tot} ——
Expected Mean Square	——	——	——	——	——	——

Remark that $SS_{tot} = SS_{GM} + SS_A + SS_B + SS_C + SS_e$. ○

For the rest of the problem, make use of the notation introduced in **F** wherever convenient, and assume normality.

G. What is the size α F-test of H_A? How does the power of this test depend on m, ν, the α_i^A's, α_j^B's, α_k^C's, and σ^2? ○

H. What are Scheffé's simultaneous confidence intervals for the family

$$\left\{ \sum_i c_i \alpha_i^A : c_1, \ldots, c_m \in \mathbb{R} \right\}?$$ ○

Suppose now that the Latin square design is enlarged through the addition of one concomitant variable, say t $(t \notin M)$: In the augmented model we have

$$E(Y_{ijk}) = \nu + \alpha_i^A + \alpha_j^B + \alpha_k^C + \gamma t_{ijk} \quad \text{for } ijk \in \mathcal{D},$$

with

$$\alpha_{\cdot}^A = \alpha_{\cdot}^B = \alpha_{\cdot}^C = 0.$$

Denote the old GME of, say, α_i^A by $\hat{\alpha}_i^A(Y) = \bar{Y}_{i..} - \bar{Y}_{...}$; write $\hat{\alpha}_i^A(t)$ for $\bar{t}_{i..} - \bar{t}_{...}$. Similarly, let $SS_e(Y)$ and $SS_e(t)$, and so on, have their obvious definitions. Use this notation to advantage in what follows.

I. In the augmented model, what are the estimators of γ, α_i^A, α_j^B, α_k^C, and σ^2? Redo parts **D**, **G**, and **H** in the context of the augmented model. $\qquad$ o

The following question should be worked after study of the next chapter.

J. Return to the original design, without the covariate. Suppose the value recorded for Y_{IJk} has been lost. What then would be the estimators of α_i^A, α_j^B, α_k^C, and σ^2? Redo parts **D**, **G**, and **H** in the context of this missing observation model. $\qquad$ o

CHAPTER 8

MISSING OBSERVATIONS

In this chapter we formulate the classical problem of missing observations in a coordinate-free framework and work out procedures for point estimation, interval estimation, and F-testing of hypotheses.

1. Framework and Gauss-Markov estimation

We begin with an example. Consider the two-way additive layout with one observation per cell: the random vector

$$Y = (Y_{ij})_{i=1,\ldots,I;\, j=1,\ldots,J} \quad \text{is weakly spherical} \tag{1.1$_1$}$$

in the space V of $I \times J$ arrays endowed with the dot-product of Exercise 2.2.26, and

$$E(Y_{ij}) = \nu + \alpha_i + \beta_j \quad \text{for } i = 1, \ldots, I, \ j = 1, \ldots, J \tag{1.1$_2$}$$

with

$$\sum_{1 \leq i \leq I} \alpha_i = 0 = \sum_{1 \leq j \leq J} \beta_j \,. \tag{1.1$_3$}$$

We presume familiarity with this model (see, for example, Exercises 2.2.26 and 4.2.14). Suppose now that Y_{IJ} is missing due to some chance effect. Perhaps its value was recorded but accidentally lost; or the experimenter never got around to observing the response to the Ith and J^{th} treatment combinations (time or money ran out); or the experimental apparatus went out of whack at this point, producing a grossly spurious value; or whatever. We assume that the loss of Y_{IJ} is not due to treatment effects (for example, death of an experimental subject due to drug overdose); the chance factors determining the loss of an observation are presumed to operate independently of the value that observation would take. Thus missingness conveys no relevant information, and, conditional on Y_{IJ} missing, (1.1) is still the appropriate model, albeit only the Y_{ij}'s with $(i, j) \neq (I, J)$ are observable.

To separate Y into its observable and unobservable components, introduce

$Y^{(1)} =$ the array Y with 0 at the site of the missing observation,

$Y^{(2)} =$ the array Y with 0's at the sites of the available observations:

$$Y^{(1)} = \left(\begin{array}{c|c} Y_{ij}\text{'s} & Y_{iJ}\text{'s} \\ \hline Y_{Ij}\text{'s} & 0 \end{array}\right)^{I \times J} \qquad Y^{(2)} = \left(\begin{array}{c|c} 0 & 0 \\ \hline 0 & Y_{IJ} \end{array}\right)^{I \times J}$$

$$\text{observable} \qquad\qquad\qquad \text{unobservable}$$

$Y^{(1)}$ and $Y^{(2)}$ take values in the subspaces

$$V^{(1)} = \{v \in V : v_{IJ} = 0\} \text{ and } V^{(2)} = \{v \in V : v_{ij} = 0 \text{ for } (i,j) \neq (I,J)\},$$

respectively. Note that

$$V^{(1)} + V^{(2)}$$

is an orthogonal decomposition of V and that

$$Y^{(1)} = P_{V^{(1)}} Y, \quad Y^{(2)} = P_{V^{(2)}} Y, \quad \text{and} \quad Y = Y^{(1)} + Y^{(2)}.$$

The observable vector $Y^{(1)}$ is weakly spherical in $V^{(1)}$ and has regression manifold

$$M^{(1)} \equiv \{ P_{V^{(1)}} m : m \in M \} = P_{V^{(1)}}(M),$$

where M is the regression manifold for Y. Thus $Y^{(1)}$ falls under the umbrella of the GLM; in particular, the Gauss-Markov estimator of its mean

$$\mu^{(1)} \equiv E\big(Y^{(1)}\big) = E(P_{V^{(1)}} Y) = P_{V^{(1)}}(EY) = P_{V^{(1)}}(\mu)$$

is just

$$\hat{\mu}^{(1)} = P_{M^{(1)}}\big(Y^{(1)}\big).$$

The point now is that μ and $\mu^{(1)}$ are in fact one-to-one functions of one another, for $P_{V^{(1)}}$ is nonsingular on M. This assertion is equivalent to the null space of $P_{V^{(1)}}$, that is, $V^{(2)}$, being disjoint from M, or to

$$\Delta_{IJ} \equiv \left(\begin{array}{c|c} 0 & 0 \\ \hline 0 & 1 \end{array}\right)^{I \times J} \notin M. \qquad (1.2)$$

(1.2) holds because, for example, the differences $(\Delta_{IJ})_{I-1,j} - (\Delta_{IJ})_{I,j}$, $1 \leq j \leq J$, do not all have the same value. The natural estimator of μ is thus

$$\hat{\mu} = (P_{V^{(1)}} | M)^{-1}(\hat{\mu}^{(1)}),$$

where $P_{V^{(1)}} | M$ denotes the restriction of $P_{V^{(1)}}$ to M. $\hat{\mu}$ gives rise to estimates of ν, α_i, and β_j in the obvious way. The objective is to determine $\hat{\mu}$ in the easiest possible way, given what one already knows about the analysis of the model (1.1) with no missing observations.

With this example as motivation, we proceed to describe what we will call the *incomplete observations model*. Suppose then that

$$Y \text{ is weakly spherical in } V, \qquad (1.3_1)$$

where $(V, \langle \cdot, \cdot \rangle)$ is an arbitrary inner product space, and

$$\mu = E(Y) \in M, \tag{1.3_2}$$

with M being a given manifold of V. Suppose also that one already knows how to treat the GLM (1.3) — in particular, P_M and Q_M are known operators.

Let Y be incompletely observed, in the following manner. V is split into two orthogonal manifolds $V^{(1)}$ and $V^{(2)}$:

$$V = V^{(1)} + V^{(2)} \quad \text{with} \quad V^{(1)} \perp V^{(2)}. \tag{1.4}$$

Set

$$Y^{(1)} = P_{V^{(1)}} Y \quad \text{and} \quad Y^{(2)} = P_{V^{(2)}} Y. \tag{1.5}$$

The interpretation is that only $Y^{(1)}$ is observable. As in the example, we assume implicitly that the unobservability of $Y^{(2)}$ does not vitiate the distributional assumption (1.3) about Y.

The classical *missing observations model* is obtained by specializing to $V = \mathbb{R}^n$ endowed with the dot-product and taking $V^{(1)}$ (respectively, $V^{(2)}$) to be the subspace of vectors with 0's at the sites of the missing (respectively, available) observations. The construct also covers what is sometimes called the *missing observations model* or the *mixed-up observations model*, an example of which is the following. Take $V = \mathbb{R}^n$, write $Y^T = (Y_1, \ldots, Y_n)$, and suppose of Y_1, Y_2, and Y_3 that we know only their sum or, equivalently, their average, A. This situation falls under the general scheme with

$$V^{(1)} = \{ v \in V : v_1 = v_2 = v_3 \}$$
$$V^{(2)} = \{ v \in V : v_1 + v_2 + v_3 = 0, v_4 = \cdots = v_n = 0 \},$$

corresponding to

$$Y^{(1)} = (A, A, A, Y_4, \ldots, Y_n)^T, \; Y^{(2)} = (Y_1 - A, Y_2 - A, Y_3 - A, 0, \ldots, 0)^T.$$

In general, the observable vector $Y^{(1)} = P_{V^{(1)}} Y$ is weakly spherical in $V^{(1)}$ with mean

$$\mu^{(1)} \equiv E(Y^{(1)}) = E(P_{V^{(1)}} Y) = P_{V^{(1)}}(EY) = P_{V^{(1)}}(\mu) \tag{1.6}$$

lying in the regression manifold

$$M^{(1)} = \{ P_{V^{(1)}} m : m \in M \} = P_{V^{(1)}}(M). \tag{1.7}$$

The Gauss-Markov estimator of $\mu^{(1)}$ is of course just

$$\hat{\mu}^{(1)} = P_{M^{(1)}} Y^{(1)}. \tag{1.8}$$

As suggested by the ANOVA example, we now assume that $\mu^{(1)}$ and μ are one-to-one functions of one another, that is, that

$$P_{V^{(1)}} \text{ is nonsingular on } M \tag{1.9_1}$$

or, equivalently,

$$\dim(M) = \dim(M^{(1)}) \tag{1.9_2}$$

or, equivalently,

$$M \text{ and } V^{(2)} \text{ are disjoint.} \tag{1.9_3}$$

We take as the estimator of μ the unique vector $\hat{\mu} \in M$ such that

$$\hat{\mu}^{(1)} = P_{V^{(1)}}(\hat{\mu}), \tag{1.10}$$

as illustrated below:

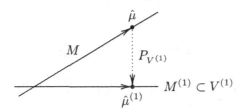

This is the right way to estimate μ, for if ψ is any linear functional on M, then $\psi(\hat{\mu})$ has minimum variance among all linear unbiased estimators of $\psi(\mu)$ of the form $\langle x, Y^{(1)} \rangle$ with $x \in V$; it is therefore appropriate to call $\hat{\mu}$ the Gauss-Markov estimator of μ.

1.11 Exercise [2]. Prove the above-stated version of the Gauss-Markov theorem for the incomplete observation model.
[Hint: Apply the usual version of the Gauss-Markov theorem to the linear functional $\psi(P_{V^{(1)}}|M)^{-1}$ in the context of the model $(V^{(1)}, Y^{(1)}, M^{(1)})$.]   ◇

The problem before us is to obtain $\hat{\mu}$ as easily as possible, building upon prior knowledge of the analysis of the GLM (1.3) with no unobservable components, in particular on knowledge of P_M. Before taking this up in detail in the next section we introduce some more notation.

For $m \in M$ we write

$$m^{(1)} = P_{V^{(1)}}m \quad \text{and} \quad m^{(2)} = P_{V^{(2)}}m, \tag{1.12}$$

so

$$m = P_{V^{(1)}}m + P_{V^{(2)}}m = m^{(1)} + m^{(2)} \tag{1.13}$$

is the unique representation of m as the sum of a vector in $M^{(1)} \subset V^{(1)}$ and a vector in the orthogonal manifold $V^{(2)}$. By (1.9), the map

$$m \to m^{(1)}$$

is invertible; we seek a better understanding of the inverse map. For this recall that, by assumption, M and $V^{(2)}$ are disjoint; let

$$\tilde{M} = M + V^{(2)}$$

be their direct sum. If $m^{(1)}$ is an arbitrary vector in $M^{(1)}$, and if m is the unique vector in M such that

$$m^{(1)} = P_{V^{(1)}} m$$

(the notation being consistent with (1.12)), then from

$$m^{(1)} = m + (-m^{(2)})$$

it follows that

$$M^{(1)} \subset \tilde{M}$$

and

$$m = P_{M;V^{(2)}} (m^{(1)}) \equiv Am^{(1)}, \tag{1.14}$$

where $A = P_{M;V^{(2)}}$ is projection onto M along $V^{(2)}$ within $\tilde{M}$. Similarly,

$$m^{(2)} = -P_{V^{(2)};M} (m^{(1)}) \equiv -Bm^{(1)}, \tag{1.15}$$

where $B = P_{V^{(2)};M}$ is projection onto $V^{(2)}$ along M within $\tilde{M}$.

1.16 Exercise [1]. Show that

$$\tilde{M} = M^{(1)} + V^{(2)},$$

with the sum being orthogonal and direct.   ◇

The following figures illustrate some of the relationships between the various quantities in the preceding discussion. The first shows the relationship between m, $m^{(1)}$, and $m^{(2)}$, viewed within the plane $\tilde{M} = M + V^{(2)} = M^{(1)} + V^{(2)}$.

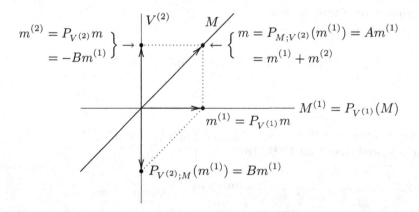

The next figure shows the relationship between $Y^{(1)}$, $\hat{\mu}^{(1)}$, $\hat{\mu}$, and $\hat{\mu}^{(2)} = P_{V^{(2)}}\hat{\mu}$, viewed within V. The subspace $V^{(1)}$ recedes back into the page, passing through the horizontal axis, the vector $Y^{(1)}$, and the subspace $M^{(1)}$. The construction ⌐ is used to indicate perpendicularity.

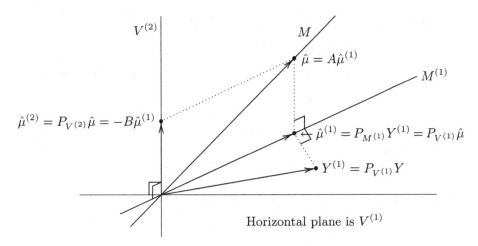

2. Obtaining $\hat{\mu}$

In the notation of the preceding section,

$$\hat{\mu}^{(1)} = P_{M^{(1)}} Y^{(1)} \tag{2.1_1}$$

is the GME of the mean $\mu^{(1)}$ of the observable component $Y^{(1)}$ of Y,

$$\hat{\mu} = P_{M;V^{(2)}}(\hat{\mu}^{(1)}) = A\hat{\mu}^{(1)} \tag{2.1_2}$$

is the GME of the mean $\mu = A\mu^{(1)}$ of Y, and

$$\hat{\mu}^{(2)} = P_{V^{(2)}}\hat{\mu} = -P_{V^{(2)};M}(\hat{\mu}^{(1)}) = \hat{\mu} - \hat{\mu}^{(1)} \tag{2.1_3}$$

is the GME of the mean of the unobservable component $Y^{(2)}$ of Y. The only operator we know explicitly is P_M, and we seek ways to determine these GMEs, especially $\hat{\mu}$, in terms of P_M and something easy to calculate. We will describe three approaches to this problem — the "consistency equation" method, the "quadratic form" method, and the "analysis of covariance" method. Although these approaches all lead to essentially the same conclusion, each sheds a somewhat different light on the solution.

2A. The consistency equation method

$\hat{\mu}^{(2)}$ estimates the mean of the unobservable $Y^{(2)}$. If we could somehow find $\hat{\mu}^{(2)}$, it might seem reasonable to use it as a substitute for the missing $Y^{(2)}$ and estimate μ by

$$P_M(Y^{(1)} + \hat{\mu}^{(2)}).$$

In fact, this is $\hat{\mu}$ for

$$
\begin{aligned}
P_M(Y^{(1)} + \hat{\mu}^{(2)}) &= P_M(\hat{\mu}^{(1)} + \hat{\mu}^{(2)}) + P_M(Y^{(1)} - \hat{\mu}^{(1)}) \\
&= P_M\hat{\mu} + P_M(P_{M^{(1)}} + P_{V^{(2)}})(Y^{(1)} - \hat{\mu}^{(1)}) \\
&= \hat{\mu} + P_M P_{M^{(1)}}(Y^{(1)} - \hat{\mu}^{(1)}) + P_M P_{V^{(2)}}(Y^{(1)} - \hat{\mu}^{(1)}) \\
&= \hat{\mu} + 0 + 0 = \hat{\mu}
\end{aligned}
$$

since $M \subset M^{(1)} + V^{(2)}$, $\hat{\mu}^{(1)} = P_{M^{(1)}}Y^{(1)}$, and $Y^{(1)} - \hat{\mu}^{(1)} \in V^{(1)}$ is orthogonal to $V^{(2)}$. Now, from

$$
\hat{\mu}^{(1)} + \hat{\mu}^{(2)} = \hat{\mu} = P_M(Y^{(1)} + \hat{\mu}^{(2)}),
$$

one obtains the so-called the *consistency equation*

$$
\hat{\mu}^{(2)} = P_{V^{(2)}} P_M(Y^{(1)} + \hat{\mu}^{(2)}).
$$

Heuristically, having decided to estimate μ by $P_M(Y^{(1)} + \hat{\mu}^{(2)})$, one ought to estimate the $V^{(2)}$-component of μ by the $V^{(2)}$-component of this estimator.

We have just seen that $m \equiv \hat{\mu}^{(2)}$ satisfies the equation

$$
m = P_{V^{(2)}} P_M(Y^{(1)} + m).
$$

We claim that this equation has a unique solution for $m \in V^{(2)}$ (or even for $m \in V$, for that matter). To see this, suppose

$$
\begin{aligned}
x &= P_{V^{(2)}} P_M(Y^{(1)} + x) \\
y &= P_{V^{(2)}} P_M(Y^{(1)} + y)
\end{aligned}
$$

for vectors $x, y \in V$, necessarily in $V^{(2)}$. Then the difference $d = y - x$ satisfies

$$
d = P_{V^{(2)}} P_M d,
$$

whence

$$
\|d\| = \|P_{V^{(2)}} P_M d\| \le \|P_M d\| \le \|d\|.
$$

The fact that equality holds at the last step forces $d \in M$. Since $d \in V^{(2)}$ as well, we must have $d = 0$.

In summary, we have

2.2 Theorem. *The unique solution m in $V^{(2)}$ (or in V) to the equation*

$$
m = P_{V^{(2)}} P_M(Y^{(1)} + m) \tag{2.3}
$$

is

$$
m = \hat{\mu}^{(2)};
$$

moreover,

$$
\hat{\mu} = P_M(Y^{(1)} + \hat{\mu}^{(2)}). \tag{2.4}
$$

This solves the estimation problem in terms of P_M, presumed to be at hand, and $P_{V^{(2)}}$, which is often very simple. For example, in the classical missing observations problem, $P_{V^{(2)}}$ is just projection onto the coordinate plane determined by the sites of the missing observations.

2.5 Exercise [2]. Suppose that a sequence $m_0, m_1, m_2, \ldots$ of approximations to $\hat{\mu}^{(2)}$ is defined as follows: $m_0 \in V^{(2)}$ is arbitrary and

$$m_n = P_{V^{(2)}} P_M(Y^{(1)} + m_{n-1})$$

for $n \geq 1$. Show that there exists a real number $c < 1$ such that

$$\|m_n - \hat{\mu}^{(2)}\| \leq c^n \|m_0 - \hat{\mu}^{(2)}\|$$

for all n.

[Hint: See Exercise 2.1.24.]   ◇

2.6 Example. Consider the two-way additive layout with one observation per cell and with Y_{IJ} missing. One has

$$(P_M y)_{ij} = \bar{y}_{i\cdot} + \bar{y}_{\cdot j} - \bar{y}_{\cdot\cdot} \tag{2.7}$$

for any vector $y = (y_{ij})_{i=1,\ldots,I, j=1,\ldots,J} \in V$ and

$$V^{(2)} = [\Delta_{IJ}] \quad \text{with} \quad \Delta_{IJ} = \left(\begin{array}{c|c} 0 & 0 \\ \hline 0 & 1 \end{array}\right)^{I \times J}.$$

According to the theorem,

$$\hat{\mu}^{(2)} = C\,\Delta_{IJ}$$

($C \in \mathbb{R}$) is determined by the consistency equation

$$\hat{\mu}^{(2)} = P_{V^{(2)}} P_M(Y^{(1)} + \hat{\mu}^{(2)})$$

or (looking only at the nonzero components of these vectors)

$$C = (\hat{\mu}^{(2)})_{IJ} = \left(P_{V^{(2)}} P_M(Y^{(1)} + C\Delta_{IJ})\right)_{IJ} = \left(P_M(Y^{(1)} + C\Delta_{IJ})\right)_{IJ}$$
$$= \frac{Y_{I\oplus} + C}{J} + \frac{Y_{\oplus J} + C}{I} - \frac{Y_{\oplus\oplus} + C}{IJ}, \tag{2.8}$$

where $\oplus$ denotes summation over all the corresponding nonmissing subscript(s). Thus

$$C\left(1 - \frac{1}{J} - \frac{1}{I} + \frac{1}{IJ}\right) = \left(\frac{Y_{I\oplus}}{J} + \frac{Y_{\oplus J}}{I} - \frac{Y_{\oplus\oplus}}{IJ}\right)$$

so

$$C = \frac{1}{(1 - 1/I)(1 - 1/J)}\left(\frac{Y_{I\oplus}}{J} + \frac{Y_{\oplus J}}{I} - \frac{Y_{\oplus\oplus}}{IJ}\right)$$
$$= \frac{Y_{I\oplus}}{J-1} + \frac{Y_{\oplus J}}{I-1} - \frac{\sum_{1 \leq i \leq I-1, 1 \leq j \leq J-1} Y_{ij}}{(I-1)(J-1)}. \tag{2.9}$$

Having obtained $\hat{\mu}^{(2)} = C\Delta_{IJ}$, one now gets $\hat{\mu} = P_M(Y^{(1)} + \hat{\mu}^{(2)})$ from (2.7), and thence $\hat{\nu}$, $\hat{\alpha}_i$, and $\hat{\beta}_j$.   •

2.10 Exercise [3]. Work out explicit formulas for $\hat{\nu}$, $\hat{\alpha}_1, \ldots \hat{\alpha}_I$, and $\hat{\beta}_1, \ldots, \hat{\beta}_J$.   ◇

2B. The quadratic function method

If we knew $Y^{(2)}$, the GME of μ would be

$$\hat{\hat{\mu}} = P_M(Y^{(1)} + Y^{(2)}), \tag{2.11}$$

the unique $m \in M$ minimizing $\|Y^{(1)} + Y^{(2)} - m\|^2$; the minimum is

$$SS_e = \|Q_M(Y^{(1)} + Y^{(2)})\|^2.$$

But as we do not know $Y^{(2)}$, it might be reasonable to use in (2.11) a fake $Y^{(2)}$ chosen so as to minimize SS_e. Under normal theory this amounts to maximizing the likelihood function not only with respect to the parameters of the problem but also the missing $Y^{(2)}$.

Let then $m^{(2)}$ be a variable point in $V^{(2)}$ and write

$$\begin{aligned}
SS_e(m^{(2)}) &\equiv \|Q_M(Y^{(1)} + m^{(2)})\|^2 \\
&= \|Y^{(1)} + m^{(2)} - P_M(Y^{(1)} + m^{(2)})\|^2 \\
&= \|Y^{(1)} - P_{V^{(1)}}P_M(Y^{(1)} + m^{(2)})\|^2 \\
&\quad + \|m^{(2)} - P_{V^{(2)}}P_M(Y^{(1)} + m^{(2)})\|^2 \\
&= \|Y^{(1)} - \hat{\mu}^{(1)}\|^2 + \|\hat{\mu}^{(1)} - P_{V^{(1)}}P_M(Y^{(1)} + m^{(2)})\|^2 \\
&\quad + \|m^{(2)} - P_{V^{(2)}}P_M(Y^{(1)} + m^{(2)})\|^2 \tag{$*$} \\
&\geq \|Q_{M^{(1)}}Y^{(1)}\|^2 + 0 + 0 = \|Q_{M^{(1)}}Y^{(1)}\|^2. \tag{2.12}
\end{aligned}$$

If we take $m^{(2)} = \hat{\mu}^{(2)}$, then, by (2.4), $P_M(Y^{(1)} + m^{(2)}) = \hat{\mu}$ and the second and third terms of ($*$) vanish, giving equality throughout (2.12). Conversely, if equality holds throughout (2.12), then $m^{(2)}$ must satisfy the consistency equation (2.3), and so be $\hat{\mu}^{(2)}$. Thus

2.13 Theorem. *The unique vector $m^{(2)}$ in $V^{(2)}$ that minimizes the quadratic function*

$$SS_e(m^{(2)}) = \|Q_M(Y^{(1)} + m^{(2)})\|^2 \tag{2.14}$$

is

$$m^{(2)} = \hat{\mu}^{(2)}.$$

Moreover,

$$\begin{aligned}
\min\{\, SS_e(m^{(2)}) : m^{(2)} \in V^{(2)} \,\} &= \|Q_M(Y^{(1)} + \hat{\mu}^{(2)})\|^2 \\
&= \|Q_{M^{(1)}}Y^{(1)}\|^2. \tag{2.15}
\end{aligned}$$

Equation (2.15) follows from (2.12) and admits the following interpretation. Having agreed to estimate μ by $P_M(Y^{(1)} + \hat{\mu}^{(2)})$, where $\hat{\mu}^{(2)}$ is the fake $Y^{(2)}$ minimizing SS_e, the minimum SS_e is in fact the sum of squares for error in the bona fide GLM involving $Y^{(1)}$ weakly spherical in $V^{(1)}$ with regression manifold $M^{(1)}$.

2.16 Example. Consider again our running example. For any $y \in V$, one has

$$\|Q_M y\|^2 = \sum_{ij} (y_{ij} - \bar{y}_{i\cdot} - \bar{y}_{\cdot j} + \bar{y}_{\cdot\cdot})^2$$
$$= \sum_{ij} y_{ij}^2 - J \sum_i \bar{y}_{i\cdot}^2 - I \sum_j \bar{y}_{\cdot j}^2 + IJ\bar{y}_{\cdot\cdot}^2.$$

If we write $m^{(2)} \in V^{(2)}$ as $m^{(2)} = C\Delta_{IJ}$, we have

$$SS_e(m^{(2)}) = \left[\left(\sum_{(i,j)\neq(I,J)} Y_{ij}^2\right) + C^2\right] - J\left[\sum_{i\neq I} \bar{Y}_{i\cdot}^2 + \left(\frac{Y_{I\oplus} + C}{J}\right)^2\right]$$
$$- I\left[\sum_{j\neq J} \bar{Y}_{\cdot j}^2 + \left(\frac{Y_{\oplus J} + C}{I}\right)^2\right] + IJ\left(\frac{Y_{\oplus\oplus} + C}{IJ}\right)^2.$$

The value of C minimizing this quadratic function is found by differentiating with respect to C and equating to 0 to get

$$C - \frac{Y_{I\oplus} + C}{J} - \frac{Y_{\oplus J} + C}{I} + \frac{Y_{\oplus\oplus} + C}{IJ} = 0.$$

Since this equation is the same as (2.8), C is again given by (2.9). ●

2C. The analysis of covariance method

This approach begins with the observation that

$$P_{M+V^{(2)}}(Y^{(1)}) = P_{M^{(1)}+V^{(2)}}(Y^{(1)})$$
$$= P_{M^{(1)}}(Y^{(1)}) + P_{V^{(2)}}(Y^{(1)})$$
$$= \hat{\mu}^{(1)} + 0 = \hat{\mu}^{(1)}. \qquad (2.17)$$

Hence, by (2.1_2) and (2.1_3),

$$\hat{\mu} = P_{M;V^{(2)}}(\hat{\mu}^{(1)}) = P_{M;V^{(2)}}\left(P_{M+V^{(2)}}(Y^{(1)})\right) \equiv \hat{\rho}_M$$
$$-\hat{\mu}^{(2)} = P_{V^{(2)};M}(\hat{\mu}^{(1)}) = P_{V^{(2)};M}\left(P_{M+V^{(2)}}(Y^{(1)})\right) \equiv \hat{\rho}_N$$

as illustrated below:

$N = V^{(2)}$

$\hat{\rho}_N = -\hat{\mu}^{(2)}$

$\hat{\mu}^{(1)} = P_{M+V^{(2)}}(Y^{(1)})$

Plane is $\tilde{M} = M + V^{(2)}$

$\hat{\rho}_M = \hat{\mu}$ M

That figure looks like the schematic diagram for analysis of covariance and so suggests that one think of $\hat{\rho}_M$ and $\hat{\rho}_N$ as the Gauss-Markov estimators of ρ_M and ρ_N in the analysis of covariance model

$$Z \text{ weakly spherical in } V \tag{2.18$_1$}$$

$$E(Z) \equiv \rho = \rho_M + \rho_N \in M + N \equiv \tilde{M}, \tag{2.18$_2$}$$

where

$$Z = Y^{(1)} \quad \text{and} \quad N = V^{(2)}. \tag{2.18$_3$}$$

The model is only formal because: (i) $Y^{(1)}$, although weakly spherical in $V^{(1)}$, is not weakly spherical in $V = V^{(1)} + V^{(2)}$, and (ii) $E(Y^{(1)}) = \mu^{(1)}$ is constrained to lie in the proper subspace $M^{(1)}$ of $\tilde{M} = M^{(1)} + V^{(2)}$.

Nonetheless, one can use the computational formulas of the analysis of covariance to determine $\hat{\mu}$ and $\hat{\mu}^{(2)}$. For example, the analysis of covariance recipe

new GME of ρ_M = old GME − old GME applied to $\hat{\rho}_N$

reads

$$\hat{\mu} = P_M Y^{(1)} - P_M \hat{\rho}_N = P_M Y^{(1)} + P_M \hat{\mu}^{(2)} = P_M (Y^{(1)} + \hat{\mu}^{(2)}),$$

which is just (2.4) of the consistency equation method. Moreover, we know how to obtain

$$\hat{\mu}^{(2)} = -\hat{\rho}_N = -B P_{\tilde{M}-M}(Y^{(1)})$$

($B = P_{N;M}$ with $N = V^{(2)}$) easily when $z_1, \ldots, z_h$ is a basis for $V^{(2)}$, to wit,

$$\hat{\mu}^{(2)} = -\sum_{1 \leq j \leq h} \hat{\gamma}_j z_j$$

with

$$\hat{\gamma} = \langle Q_M z, Q_M z \rangle^{-1} \langle Q_M z, Y^{(1)} \rangle.$$

In the classical case of missing observations in $\mathbb{R}^n$, a natural choice for a basis of $N = V^{(2)}$ consists of the $\dim(V^{(2)})$ covariate vectors, one for each missing observation, each consisting of all 0's except for a 1 at the site of the corresponding missing observation. Remember that in this case $Y^{(1)}$ has 0's at the sites of the missing observations.

2.19 Example. Consider again our running example, where $V^{(2)} = [\Delta_{IJ}]$. With $z = \Delta_{IJ}$, we find

$$\|Q_M z\|^2 = \langle Q_M z, z \rangle = 1 - \frac{1}{J} - \frac{1}{I} + \frac{1}{IJ} = (1 - 1/I)(1 - 1/J) \tag{2.20}$$

and

$$\langle Q_M z, Y^{(1)} \rangle = \langle z, Y^{(1)} \rangle - \langle P_M z, Y^{(1)} \rangle = 0 - \langle z, P_M Y^{(1)} \rangle$$

$$= -(P_M Y^{(1)})_{IJ} = -\left(\frac{1}{J} Y_{I\oplus} + \frac{1}{I} Y_{\oplus J} - \frac{1}{IJ} Y_{\oplus\oplus} \right).$$

Since the resulting equation $\langle Q_M z, Q_M z \rangle C = \langle Q_M z, Y^{(1)} \rangle$ for $C = -\hat{\gamma}$ is the same as (2.8),

$$\hat{\mu}^{(2)} = -\hat{\gamma} z = -\hat{\gamma} \Delta_{IJ}$$

is as before (see again (2.9)).   •

2.21 Exercise [3]. Suppose $z_1, \ldots, z_h$ is a basis for $N = V^{(2)}$. Let $x^{(1)}, \ldots, x^{(h)}$ be vectors in N such that

$$\langle z_i, x^{(j)} \rangle = \delta_{ij} \quad \text{for } i, j = 1, \ldots, h;$$

$x^{(j)}$ is the coefficient vector for the functional $\sum_i c_i z_i \to c_j$. Show that

$$m = \sum_{1 \leq j \leq h} c_j z_j \in N$$

satisfies the consistency equation

$$m = P_{V^{(2)}} P_M (Y^{(1)} + m)$$

if and only if

$$\langle x^{(i)}, m \rangle = \langle x^{(i)}, P_{V^{(2)}} P_M (Y^{(1)} + m) \rangle \quad \text{for } i = 1, \ldots, h, \qquad (2.22_1)$$

if and only if

$$-\langle x^{(i)}, Q_M Y^{(1)} \rangle = \sum_j \langle x^{(i)}, Q_M z_j \rangle c_j \qquad \text{for } i = 1, \ldots, h, \qquad (2.22_2)$$

if and only if

$$-\langle z_i, Q_M Y^{(1)} \rangle = \sum_j \langle z_i, Q_M z_j \rangle c_j \qquad \text{for } i = 1, \ldots, h, \qquad (2.22_3)$$

and if and only if

$$-\langle Q_M z_i, Y^{(1)} \rangle = \sum_j \langle Q_M z_i, Q_M z_j \rangle c_j \qquad \text{for } 1 = 1, \ldots, h, \qquad (2.22_4)$$

thereby verifying anew the relation $\hat{\mu}^{(2)} = -\hat{\rho}_N$ via the consistency equation characterization of $\hat{\mu}^{(2)}$ and the basis characterization of $\hat{\rho}_N$.

(Note that in the classical missing observations model with $z_1, \ldots, z_h$ the usual orthonormal basis for $V^{(2)}$, one has $x^{(i)} = z_i$ for each i, and (2.22_1) merely expresses the consistency equation in coordinate form.)   ◇

2.23 Exercise [4]. Consider again the two-way layout with one observation per cell, but suppose now that $Y_{I-1,J-1}$, $Y_{I-1,J}$, $Y_{I,J-1}$, and Y_{IJ} are each missing. Assume I and J are both 3 or more. Show, using any of the three methods discussed above, that the nonzero components of $\hat{\mu}^{(2)}$ are

$$(\hat{\mu}^{(2)})_{ij} = \frac{Y_{i\oplus}}{J-2} + \frac{Y_{\oplus j}}{I-2} - \frac{\sum_{1 \leq i \leq I-2} \sum_{1 \leq j \leq J-2} Y_{ij}}{(I-2)(J-2)} \qquad (2.24)$$

for $i = I - 1, I$ and $j = J - 1, J$. Check (2.24) using the guessing method, which says that $\langle m^{(1)}, Y^{(1)} \rangle$ is the Gauss-Markov estimator of its expected value, provided $m^{(1)} \in M^{(1)}$. [Hint: If $\begin{pmatrix} a & b \\ c & d \end{pmatrix} = P_{V^{(2)}} P_M y$, then $a + d = b + c$.]   ◇

3. Estimation of σ^2

By assumption, the dispersion operator of Y is $\sigma^2 I_V$. The dispersion operator of $Y^{(1)}$ is thus $\sigma^2 I_{V^{(1)}}$, and since $Y^{(1)}$ has regression manifold $M^{(1)}$, the "proper" estimator of σ^2 is

$$\hat{\sigma}^2 = \frac{SS_e}{d(V^{(1)} - M^{(1)})}, \tag{3.1}$$

where (in contrast to its usage in Section 2B above) SS_e now denotes $\|Q_{M^{(1)}} Y^{(1)}\|^2$. By (1.9) one has

$$
\begin{aligned}
d(V^{(1)} - M^{(1)}) &= d(V^{(1)}) - d(M^{(1)}) = d(V^{(1)}) - d(M) \\
&= d(V) - d(M) - d(V^{(2)}) \\
&= d(M^\perp) - d(V^{(2)}) \\
&= \text{old degrees of freedom} \\
&\quad - \text{"dimension" of the unobservable } Y^{(2)}.
\end{aligned} \tag{3.2}
$$

This is exactly the reduction in the degrees of freedom (d.f.) for the formal analysis of covariance model (2.18):

$$
\begin{aligned}
\text{new d.f.} &= \text{old d.f.} - \text{dimension of the covariate manifold} \\
&= \text{old d.f.} - d(N) = \text{old d.f.} - d(V^{(2)}).
\end{aligned}
$$

As for SS_e, from the formulas in the consistency equation method, one has

$$
\begin{aligned}
SS_e &= \left\| Q_{M^{(1)}} Y^{(1)} \right\|^2 = \left\| Y^{(1)} - \hat{\mu}^{(1)} \right\|^2 \\
&= \left\| Y^{(1)} + \hat{\mu}^{(2)} - (\hat{\mu}^{(1)} + \hat{\mu}^{(2)}) \right\|^2 = \left\| Y^{(1)} + \hat{\mu}^{(2)} - \hat{\mu} \right\|^2 \\
&= \left\| Y^{(1)} + \hat{\mu}^{(2)} - P_M(Y^{(1)} + \hat{\mu}^{(2)}) \right\|^2 = \left\| Q_M(Y^{(1)} + \hat{\mu}^{(2)}) \right\|^2.
\end{aligned} \tag{3.3}
$$

Thus, if one uses $Y^{(1)} + \hat{\mu}^{(2)}$ as a fake Y, then the associated sum of squares for error is in fact the proper SS_e. We discovered this recipe for SS_e earlier in connection with the quadratic function method (see Theorem 2.13, especially (2.15)).

The formal analysis of covariance model (2.18) also gives a valid formula for SS_e. Indeed, from (2.17), that is,

$$\hat{\mu}^{(1)} = P_{\tilde{M}} Y^{(1)}$$

with $\tilde{M} = M + V^{(2)} = M + N$, there follows

$$
\begin{aligned}
\text{ANCOVA } SS_e &\equiv \left\| Q_{\tilde{M}} Y^{(1)} \right\|^2 = \left\| Y^{(1)} - P_{\tilde{M}} Y^{(1)} \right\|^2 \\
&= \left\| Y^{(1)} - \hat{\mu}^{(1)} \right\|^2 = \left\| Q_{M^{(1)}} Y^{(1)} \right\|^2.
\end{aligned}
$$

SS_e may therefore be determined from the analysis of covariance computing formula

$$\text{new } SS = \text{old } SS - \text{adjustment} = \left\| Q_M Y^{(1)} \right\|^2 - \left\| P_{\tilde{M}-M} Y^{(1)} \right\|^2, \tag{3.4}$$

with

$$\left\|P_{\tilde{M}-M}Y^{(1)}\right\|^2 = \hat{\gamma}^T \langle Q_M z, Q_M z\rangle \hat{\gamma}$$
$$= \langle Q_M z, Y^{(1)}\rangle^T \langle Q_M z, Q_M z\rangle^{-1} \langle Q_M z, Y^{(1)}\rangle$$

and

$$\hat{\gamma} = \langle Q_M z, Q_M z\rangle^{-1} \langle Q_M z, Y^{(1)}\rangle$$

in the basis case.

4. *F*-testing

Suppose in the unobservable components model we wish to test

$$H: \mu \in M_0 \quad \text{versus} \quad A: \mu \notin M_0, \tag{4.1}$$

where M_0 is a given subspace of M. Since $P_{V^{(1)}}$ is nonsingular on M by assumption, it is also nonsingular on M_0, and we may therefore treat

$$M_0^{(1)} \equiv P_{V^{(1)}}(M_0)$$

on the same footing as $M^{(1)}$. To (4.1) there corresponds the testing problem

$$H: \mu^{(1)} \in M_0^{(1)} \quad \text{versus} \quad A: \mu^{(1)} \notin M_0^{(1)} \tag{4.2}$$

based on $Y^{(1)}$. The appropriate *F*-statistic is

$$F = \frac{(SS_{e,H} - SS_e)/d(M^{(1)} - M_0^{(1)})}{SS_e/d(V^{(1)} - M^{(1)})} = \frac{(SS_{e,H} - SS_e)/d(M^{(1)} - M_0^{(1)})}{\hat{\sigma}^2} \tag{4.3}$$

with

$$d(M^{(1)} - M_0^{(1)}) = d(M^{(1)}) - d(M_0^{(1)}) = d(M) - d(M_0) = d(M - M_0)$$

and

$$SS_e = \left\|Q_{M^{(1)}}(Y^{(1)})\right\|^2 \quad \text{and} \quad SS_{e,H} = \left\|Q_{M_0^{(1)}}(Y^{(1)})\right\|^2.$$

These sums of squares can be obtained by any of the methods of the preceding section, in particular from the analysis of covariance formula (3.4). Since (4.2) translates into

$$H: \rho \in M_0 + V^{(2)} \quad \text{versus} \quad A: \rho \notin M_0 + V^{(2)}$$

in the formal analysis of covariance framework, that is, since the formal covariate manifold N is $V^{(2)}$ under both the general assumptions and under H, use of the standard analysis of covariance formula for the adjusted *F*-statistic gives the correct value.

4.4 Exercise [2]. In connection with the consistency equation/quadratic function methods for the calculation of the SS_e's, it is sometimes recommended for the sake of computational simplicity that the same fake $Y^{(2)} = \hat{\mu}^{(2)}$ obtained under the general assumptions be used as well under H; this is the so-called *Yates procedure*. Use Theorem 2.13 to show that the Yates procedure overestimates the true F-statistic, and discuss the implications.   ◇

Assuming $Y^{(1)}$ is normally distributed, the F-statistic (4.3) is distributed as $\mathcal{F}$ with $d(M - M_0)$ and $d(M^\perp) - d(V^{(2)})$ degrees of freedom, and noncentrality parameter δ given by

$$
\begin{aligned}
\sigma^2 \delta^2 &= \left\| Q_{M_0^{(1)}} \mu^{(1)} \right\|^2 \\
&= \left\| Q_{\tilde{M}_0} \mu^{(1)} \right\|^2 && (\tilde{M}_0 \equiv M_0^{(1)} + V^{(2)}) \\
&= \left\| Q_{\tilde{M}_0} \mu - Q_{\tilde{M}_0} \mu^{(2)} \right\|^2 && (\mu = \mu^{(1)} + \mu^{(2)}) \\
&= \left\| Q_{\tilde{M}_0} \mu \right\|^2 && (\mu^{(2)} \in V^{(2)} \subset \tilde{M}_0) \\
&= \left\| Q_{M_0} \mu \right\|^2 - \left\| P_{\tilde{M}_0 - M_0} \mu \right\|^2 \\
&= \text{old formula} - \text{correction term} && (4.5)
\end{aligned}
$$

with

$$
\text{correction term} = \langle Q_{M_0} z, Q_{M_0} \mu \rangle^T \langle Q_{M_0} z, Q_{M_0} z \rangle^{-1} \langle Q_{M_0} z, Q_{M_0} \mu \rangle \qquad (4.6)
$$

in the basis case $V^{(2)} = [z_1, \ldots, z_h]$. These formulas are just specializations of the usual analysis of covariance formulas.

4.7 Example. Let us take up again our running example and consider testing the common null hypothesis of no row effect

$$
H_A: \alpha_i = 0 \quad \text{for } i = 1, \ldots, I. \qquad (4.8)
$$

Writing M_A for M_0 and using the formula

$$
(P_{M_A} y)_{ij} = \bar{y}_{\cdot j},
$$

we find $\hat{\mu}_{H_A}^{(2)} \equiv C_{H_A} \Delta_{IJ}$ from the consistency equation

$$
\begin{aligned}
C_{H_A} = (\hat{\mu}_{H_A}^{(2)})_{IJ} &= \left(P_{V^{(2)}} P_{M_A} (Y^{(1)} + \hat{\mu}_{H_A}^{(2)}) \right)_{IJ} \\
&= \left(P_{M_A} (Y^{(1)} + \hat{\mu}_{H_A}^{(2)}) \right)_{IJ} = \frac{Y_{\oplus J} + C_{H_A}}{I}
\end{aligned}
$$

or

$$
C_{H_A} = \frac{Y_{\oplus J}}{I - 1}. \qquad (4.9)
$$

With C_{H_A} in hand in simple form, there is now no difficulty in computing $SS_{e,H_A} = \left\| Q_{M_A}(Y^{(1)} + C_{H_A}\Delta_{IJ}) \right\|^2$ via the usual recipe

$$\|Q_{M_A} y\|^2 = \sum_{ij}(y_{ij} - \bar{y}_{\cdot j})^2 = \sum_{ij} y_{ij}^2 - I \sum_j \bar{y}_{\cdot j}^2.$$

It may be helpful to give a simple numerical example of the test of (4.8). Take $I = 2$ and $J = 3$ and suppose the sample is

i	1	2	3
1	-2	0	2
2	4	2	?

(with j labeling the columns)

where ? denotes a missing value. We will compute the F-statistic via the consistency equation approach. From (2.9) we find that

$$C = \frac{\text{estimate of the mean of the missing observation}}{\text{under the general assumptions}}$$

$$= \frac{Y_{I\oplus}}{J-1} + \frac{Y_{\oplus J}}{I-1} - \frac{\sum_{1 \le i \le I-1, 1 \le j \le J-1} Y_{ij\bullet}}{(I-1)(J-1)}$$

$$= \frac{4+2}{2} + \frac{2}{1} - \frac{-2+0}{2} = 3 + 2 - (-1) = 6.$$

The fake Y, namely "Y" $= Y^{(1)} + C\Delta_{IJ}$, is

				row averages	$\hat{\alpha}_i$
"Y" $=$	-2	0	2	0	-2
	4	2	6	4	2
column averages	1	1	4	2	
$\hat{\beta}_j$'s	-1	-1	2		$\hat{\nu} = 2$

Thus (see (Exercise 4.2.14))

$$P_M(\text{``}Y\text{''}) = \begin{pmatrix} -1 & -1 & 2 \\ 3 & 3 & 6 \end{pmatrix} \quad \text{and} \quad Q_M(\text{``}Y\text{''}) = \begin{pmatrix} -1 & +1 & 0 \\ +1 & -1 & 0 \end{pmatrix},$$

and

$$SS_e = \|Q_M(\text{``}Y\text{''})\|^2 = 4$$

with

$$(I-1)(J-1) - 1 = 1 \times 2 - 1 = 1$$

degree of freedom.

Even more simply, for testing the null hypothesis H_A of no row effects we obtain from (4.9) that

$$C_{H_A} = \text{estimate of the mean of the missing observation under } H_A$$

$$= \frac{Y_{\oplus J}}{I-1} = 2$$

(notice that the Yates approximation, namely $C_{H_A} \approx C$, is very poor here), whence the fake Y, namely "Y_{H_A}" $= Y^{(1)} + C_{H_A}\Delta_{IJ}$, under H_A is

$$\text{``}Y_{H_A}\text{''} = \begin{pmatrix} -2 & 0 & 2 \\ 4 & 2 & 2 \end{pmatrix}$$

$$column~averages \quad 1 \quad 1 \quad 2$$

so

$$P_{M_A}(\text{``}Y_{H_A}\text{''}) = \begin{pmatrix} 1 & 1 & 2 \\ 1 & 1 & 2 \end{pmatrix} \quad \text{and} \quad Q_{M_A}(\text{``}Y_{H_A}\text{''}) = \begin{pmatrix} -3 & -1 & 0 \\ 3 & 1 & 0 \end{pmatrix}.$$

Thus

$$SS_{e,H} = 9 + 9 + 1 + 1 = 20$$

with

$$J(I-1) - 1 = 3 \times 1 - 1 = 2$$

degrees of freedom.

Consequently,

$$SS_{e,H} - SS_e = 20 - 4 = 16$$

with

$$2 - 1 = 1$$

degrees of freedom, and the test statistic is

$$F = \frac{16/1}{4/1} = 4, \tag{4.10}$$

which is regarded as an $\mathcal{F}_{1,1}$ variate under the null hypothesis.

The power of the F-test in this example depends on the noncentrality parameter δ given by

$$\sigma^2\delta^2 = \text{old formula} - \text{correction}$$

$$= \|Q_{M_A}\mu\|^2 - \frac{\langle Q_{M_A}\Delta_{23}, \mu\rangle^2}{\langle Q_{M_A}\Delta_{23}, Q_{M_A}\Delta_{23}\rangle} = J\sum_i \alpha_i^2 - \frac{\langle \Delta_{23}, Q_{M_A}\mu\rangle^2}{\|Q_{M_A}\Delta_{23}\|^2}$$

$$= J\sum_i \alpha_i^2 - \frac{\alpha_I^2}{1/2} = J\sum_i \alpha_i^2 - \frac{I}{I-1}\alpha_I^2,$$

where

$$\Delta_{23} = \begin{pmatrix} 0 & 0 & 0 \\ 0 & 0 & 1 \end{pmatrix};$$

note that

$$P_{M_A}\Delta_{23} = \begin{pmatrix} 0 & 0 & 1/2 \\ 0 & 0 & 1/2 \end{pmatrix} \quad \text{and} \quad Q_{M_A}\Delta_{23} = \begin{pmatrix} 0 & 0 & -1/2 \\ 0 & 0 & +1/2 \end{pmatrix}.$$

•

4.11 Exercise [3]. Arrive at (4.10) following the analysis of covariance route.   ◇

5. Estimation of linear functionals

Suppose ψ is a linear functional on M with coefficient vector $x \in M$:

$$\psi(m) = \langle x, m \rangle \quad \text{for all } m \in M.$$

The associated functional ψ^* on $M^{(1)}$ defined by

$$\psi^*(m^{(1)}) = \psi(P_{M;V^{(2)}}m^{(1)}) = \langle x, P_{M;V^{(2)}}m^{(1)} \rangle$$
$$= \langle x, Am^{(1)} \rangle = \langle A'x, m^{(1)} \rangle$$

has by Proposition 7.1.4 coefficient vector

$$A'x = (P_{M;V^{(2)}})'x = P_{\tilde{M}-V^{(2)};\tilde{M}-M}x = P_{M^{(1)};\tilde{M}-M}x; \tag{5.1}$$

here and throughout this section, the operation of the various nonorthogonal projections involved is understood to be on the manifold

$$\tilde{M} = M + V^{(2)} = M^{(1)} + V^{(2)}.$$

We note for future reference that the map

$$\psi \to \psi^*$$

is one-to-one, because the map $x \equiv cv\,\psi \to A'x \equiv cv\,\psi^* = P_{M^{(1)};\tilde{M}-M}x$ is nonsingular on M.

  The estimator

$$\psi(\hat{\mu}) = \psi(P_{M;V^{(2)}}\hat{\mu}^{(1)}) = \psi^*(\hat{\mu}^{(1)}) = \psi^*(P_{V^{(1)}}Y^{(1)})$$

estimates $\psi(\mu)$ linearly and unbiasedly because

$$E_\mu\psi(\hat{\mu}) = E_\mu\psi^*(\hat{\mu}^{(1)}) = \psi^*(\mu^{(1)}) = \psi(\mu).$$

Moreover, if $v \in V$ and $\langle v, Y^{(1)} \rangle_V = \langle P_{V^{(1)}}v, Y^{(1)} \rangle_{V^{(1)}}$ is an unbiased estimator of $\psi(\mu) = \psi^*(\mu^{(1)})$, then by the Gauss-Markov theorem for $Y^{(1)}$, $M^{(1)}$, and ψ^*, one has $\text{Var}(\langle P_{V^{(1)}}v, Y^{(1)} \rangle_{V^{(1)}}) \geq \text{Var}(\psi^*(\hat{\mu}^{(1)}))$, or

$$\text{Var}(\langle v, Y^{(1)} \rangle) \geq \text{Var}(\psi(\hat{\mu})),$$

with equality if and only if $\langle v, Y^{(1)} \rangle = \psi(\hat{\mu})$. Thus $\psi(\hat{\mu})$ is the best linear unbiased for $\psi(\mu)$, as asserted earlier in Exercise 1.11.

Recall $B = P_{V^{(2)};M}$ and $B' = P_{\tilde{M}-M;M^{(1)}}$. By (5.1),

$$\mathrm{Var}\big(\psi(\hat{\mu})\big) = \mathrm{Var}\big(\psi^*(\hat{\mu}^{(1)})\big) = \mathrm{Var}\big(\langle A'x, \hat{\mu}^{(1)}\rangle\big) = \mathrm{Var}\big(\langle A'x, Y^{(1)}\rangle\big)$$
$$= \sigma^2 \|A'x\|^2 = \sigma^2 \|x - B'x\|^2 = \sigma^2\big(\|x\|^2 + \|B'x\|^2\big)$$
$$= \text{old formula} + \text{correction}, \qquad (5.2)$$

with $x = cv\,\psi \in M$. Moreover, in the case when $z_1, \ldots, z_h$ is a basis for $V^{(2)}$, Proposition 7.1.18 gives

$$\text{correction}/\sigma^2 = \|B'x\|^2 = \langle z, x\rangle^T \langle Q_M z, Q_M z\rangle^{-1} \langle z, x\rangle$$
$$= \hat{\psi}(z)^T \langle Q_M z, Q_M z\rangle^{-1} \hat{\psi}(z), \qquad (5.3)$$

where

$$\hat{\psi}(z) = \big(\hat{\psi}(z_1), \ldots, \hat{\psi}(z_h)\big)^T,$$

with

$$\hat{\psi}(z_i) = \psi(P_M z_i) = \langle x, P_M z_i\rangle = \langle x, z_i\rangle \quad \text{for } 1 \le i \le h,$$

with the ˆ's here designating Gauss-Markov estimation in the original model, with no unobservable components. The natural estimator of $\mathrm{Var}\big(\psi(\hat{\mu})\big)$ is of course

$$\hat{\sigma}^2_{\hat{\psi}} \equiv \hat{\sigma}^2\big(\|x\|^2 + \|B'x\|^2\big) \qquad (5.4)$$

$\big(x = cv\,\psi\big)$ with

$$\hat{\sigma}^2 = \frac{\|Q_{M^{(1)}} Y^{(1)}\|^2}{d(V^{(1)} - M^{(1)})}$$

as in Section 3. Note that (5.2), (5.3), and (5.4) are just the analysis of covariance formulas associated with the model (2.18).

5.5 Example. Consider again our running example of the two-way layout with one observation per cell and Y_{IJ} missing. $V^{(2)}$ is spanned by $z = \Delta_{IJ}$. We will use (5.2) and (5.3) to calculate $\mathrm{Var}(\hat{\nu})$, $\mathrm{Var}(\hat{\alpha}_i)$, and $\mathrm{Var}(\hat{\beta}_j)$:

$$\mathrm{Var}(\hat{\nu}) = \text{old formula} + \text{correction} \qquad \text{(by (5.2))}$$

$$= \sigma^2\left[\frac{1}{IJ} + \frac{(\bar{z}_{..})^2}{\|Q_M z\|^2}\right] \qquad \text{(by (5.3), with } \hat{\nu}(z) = \bar{z}_{..})$$

$$= \sigma^2\left[\frac{1}{IJ} + \frac{1/(IJ)^2}{(1 - 1/I)(1 - 1/J)}\right] \qquad \text{(by (2.20))}$$

$$= \frac{\sigma^2}{IJ}\left[1 + \frac{1}{(I-1)(J-1)}\right] \qquad (5.6_1)$$

and

$$\text{Var}(\hat{\alpha}_i) = \text{old formula} + \text{correction}$$

$$= \sigma^2 \left[\left(\frac{1}{J} - \frac{1}{IJ} \right) + \frac{(\bar{z}_{i\cdot} - \bar{z}_{\cdot\cdot})^2}{\|Q_M z\|^2} \right]$$

$$= \sigma^2 \left[\frac{I-1}{IJ} + \frac{\left(\frac{1}{J}\delta_{iI} - \frac{1}{IJ} \right)^2}{(1 - 1/I)(1 - 1/J)} \right]$$

$$= \begin{cases} \sigma^2 \frac{I-1}{IJ} \left[1 + \frac{1}{(I-1)^2(J-1)} \right], & \text{if } i \ne I, \\ \sigma^2 \frac{I-1}{IJ} \left[1 + \frac{1}{J-1} \right], & \text{if } i = I. \end{cases} \tag{5.6$_2$}$$

Of course $\text{Var}(\hat{\beta}_j)$ is given by (5.6$_2$) with the roles of rows and columns interchanged.   •

5.7 Exercise [3]. Work out $\text{Var}(\sum_{1 \le i \le I} c_i \hat{\alpha}_i)$ for $c_1, \dots, c_I \in \mathbb{R}$.   ◇

Suppose next that $\mathcal{L}$ is a space of linear functionals on M. Put

$$\mathcal{L}^* = \{ \psi^* : \psi \in \mathcal{L} \}.$$

Under the assumption of normality, application of the Scheffé multiple comparison procedures to $\mathcal{L}^*$ in the context of the $(V^{(1)}, Y^{(1)}, M^{(1)})$ GLM gives

$$P_{\mu^{(1)}, \sigma^2} \left(|\psi^*(\hat{\mu}^{(1)}) - \psi^*(\mu^{(1)})| \le S\hat{\sigma} \| cv \, \psi^* \| \text{ for all } \psi^* \in \mathcal{L}^* \right) = 1 - \alpha$$

for all $\mu^{(1)} \in M^{(1)}$ and $\sigma^2 > 0$, with

$$S = S_{d(\mathcal{L}^*), d(V^{(1)} - M^{(1)})}(\alpha).$$

But

$$d(\mathcal{L}^*) = d(\mathcal{L}),$$

since as observed above the map $\psi \to \psi^*$ is one-to-one, and

$$\psi^*(\hat{\mu}^{(1)}) = \psi(\hat{\mu}) \quad \text{and} \quad \psi^*(\mu^{(1)}) = \psi(\mu).$$

Thus

$$P_{\mu, \sigma^2} \left(|\psi(\hat{\mu}) - \psi(\mu)| \le S\hat{\sigma}_{\hat{\psi}} \text{ for all } \psi \in \mathcal{L} \right) = 1 - \alpha \tag{5.8}$$

for all $\mu \in M$ and $\sigma^2 > 0$, with $\hat{\sigma}_{\hat{\psi}}^2$ given by (5.4) and

$$S = S_{d(\mathcal{L}), d(M^\perp) - d(V^{(2)})}(\alpha)$$

(see (3.2)).

5.9 Exercise [2]. Specialize (5.8) to the case where $\mathcal{L}$ is the collection of all linear combinations of the α_i's in the running model.   ◇

6. Problem set: Extra observations

The preceding sections have dealt with the analysis of models in which some of the observations one would customarily have at hand are missing. In contrast, this problem set deals with the analysis of models in which observations are available in addition to those customarily taken. Only material in Chapters 1 through 6 is called upon.

To state the problem abstractly, let Y be a weakly spherical random vector $\big(\Sigma(Y) = \sigma^2 I\big)$ taking values in a given inner product space $(V, \langle \cdot, \cdot \rangle)$ and having mean μ lying in a given subspace M of V. Let

$$V = V^{(1)} + V^{(2)}$$

be an orthogonal decomposition of V. For $i = 1, 2$, put

$$Y^{(i)} = P_{V^{(i)}} Y$$

and set

$$M^{(i)} = P_{V^{(i)}}(M).$$

Suppose that it is known how to treat the general linear models involving $Y^{(i)}$ weakly spherical in $V^{(i)}$ with regression manifold $M^{(i)}$ for $i = 1, 2$. The point of this problem set is to see how one can build on this knowledge in analyzing the general linear model involving Y weakly spherical in V with regression manifold M.

As an example of the setup just described, suppose $Y_1, \ldots, Y_n$ are uncorrelated random variables with common variance σ^2 and means of the form

$$E(Y_j) = \sum\nolimits_{1 \leq k \leq p} x_{jk} \beta_k \quad \text{for } 1 \leq j \leq n,$$

where the design matrix $(x_{jk})_{j=1,\ldots,n;\,k=1,\ldots,p}$ is of full rank, so one can make inferences on the treatment effects $\beta_1, \ldots, \beta_p$. Suppose now some extra observations are made on, say, the last treatment combination, that is, the previous model is augmented to include m additional random variables $Y_{n\ell}$, $\ell = 1, \ldots, m$, having variance σ^2, zero correlation among themselves and with the original Y_j's, and means

$$E(Y_{n\ell}) = \sum\nolimits_{1 \leq k \leq p} x_{nk} \beta_k \quad \text{for } 1 \leq \ell \leq m.$$

This instance of the so-called *replicated observations model* falls under the general framework above upon taking

$$V = \mathbb{R}^{n+m},$$

$$Y = (Y_1, \ldots, Y_n, Y_{n1}, \ldots, Y_{nm})^T,$$

and

$$V^{(1)} = \big\{ (y_1, \ldots, y_n, y_{n1}, \ldots, y_{nm})^T \in V : y_{n1} = \cdots = y_{nm} = 0 \big\}$$
$$V^{(2)} = \big\{ (y_1, \ldots, y_n, y_{n1}, \ldots, y_{nm})^T \in V : y_1 = \cdots = y_n = 0 \big\}.$$

The regression problem involving $Y^{(1)}$ and $M^{(1)}$ is essentially just the original problem involving $(Y_1, \ldots, Y_n)^T$ and its regression manifold — a problem we presume to be well understood — while the regression problem involving $Y^{(2)}$ and $M^{(2)}$ is essentially just the one-sample problem, involving estimation of the common mean of the extra observations $Y_{n1}, \ldots, Y_{nm}$.

Return now to the general formulation. Add to the framework already established the following two features of the particular replicated observations model above:

$$d(M^{(1)}) = d(M) \tag{6.1}$$

and

$$d(M^{(2)}) = 1. \tag{6.2}$$

Choose and fix a nonzero $x \in M^{(2)}$.

A. By (6.1), the map

$$m \to m^{(1)} = P_{V^{(1)}} m$$

defined on M is invertible. Show that there exists a unique vector $z \in M^{(1)}$ such that the inverse map is

$$m^{(1)} \to m^{(1)} + \langle z, m^{(1)} \rangle x. \tag{6.3} \circ$$

Now write

$$\mu^{(1)} = EY^{(1)} \quad \text{and} \quad \mu^{(2)} = EY^{(2)} = \gamma x;$$

by (6.3), γ and $\mu^{(1)}$ are related by

$$\gamma = \langle z, \mu^{(1)} \rangle.$$

The GME of γ based on $Y^{(1)}$ alone is

$$\hat{\gamma}^{(1)} = \langle z, Y^{(1)} \rangle,$$

while the GME of γ based on $Y^{(2)}$ alone is

$$\hat{\gamma}^{(2)} = \langle y, Y^{(2)} \rangle,$$

where

$$y = \frac{x}{\|x\|^2} .$$

Put

$$v = z - y.$$

B. Put

$$N = (M^{(1)} + M^{(2)}) - M.$$

Show that

$$d(N) = d(M^{(2)})$$

and in fact

$$N = [v]. \qquad \circ$$

C. Show that

$$P_M Y = P_{M^{(1)}} Y^{(1)} + P_{M^{(2)}} Y^{(2)} - P_N Y$$
$$= P_{M^{(1)}} Y^{(1)} + \hat{\gamma}^{(2)} x - c\hat{\delta} v$$

and

$$P_{M^{(1)}} P_M Y = P_{M^{(1)}} (Y^{(1)} - c\hat{\delta} z),$$

where

$$\hat{\delta} = \hat{\gamma}^{(1)} - \hat{\gamma}^{(2)}$$

and

$$\frac{1}{c} = \|v\|^2 = \|z\|^2 + \|y\|^2 = \frac{\text{Var}(\hat{\delta})}{\sigma^2}.$$ ○

D. Show that each linear functional ψ on M can be written uniquely in the form

$$\psi(\mu) = \psi^{(1)}(P_{M^{(1)}} \mu) = \langle w, \mu \rangle,$$

where $\psi^{(1)}$ is a linear functional on $M^{(1)}$ and $w \in M^{(1)}$ is its coefficient vector. Show also that the GME of $\psi(\mu)$ is

$$\hat{\psi}(Y) = \psi^{(1)}(P_{M^{(1)}} P_M Y) = \hat{\psi}^{(1)}(Y^{(1)}) - c\hat{\delta} \psi^{(1)}(z)$$

and that

$$\text{Var}(\hat{\psi}(Y)) = \sigma^2 \|P_M w\|^2 = \sigma^2 \big[\|w\|^2 - c(\psi^{(1)}(z))^2 \big].$$

What is the corresponding formula for the covariance between the GMEs of two linear functionals of μ?   ○

E. In connection with the customary unbiased estimator of σ^2, namely

$$\hat{\sigma}^2 = \frac{\|Q_M Y\|^2}{d(M^\perp)},$$

show that

$$d(M^\perp) = d(V^{(1)} - M^{(1)}) + d(V^{(2)})$$

and that

$$\|Q_M Y\|^2 = \left\| P_{V^{(1)} - M^{(1)}} Y^{(1)} \right\|^2 + \left\| P_{V^{(2)} - M^{(2)}} Y^{(2)} \right\|^2 + \left\| P_N Y \right\|^2$$
$$= \left\| P_{V^{(1)} - M^{(1)}} Y^{(1)} \right\|^2 + \left\| Y^{(2)} - \hat{\gamma}^{(2)} x \right\|^2 + c\hat{\delta}^2.$$ ○

F. Suppose M_0 is a subspace of M. Put

$$M_0^{(1)} = P_{V^{(1)}}(M_0), \quad M_0^{(2)} = P_{V^{(2)}}(M_0), \quad \text{and} \quad N_0 = (M_0^{(1)} + M_0^{(2)}) - M_0.$$

Show that the F-statistic for testing

$$H : \mu \in M_0 \quad \text{versus} \quad A : \mu \notin M_0$$

is

$$\frac{\left\|P_{M^{(1)}-M_0^{(1)}}Y^{(1)}\right\|^2 + \left\|P_{M^{(2)}-M_0^{(2)}}Y^{(2)}\right\|^2 + \left\|P_{N_0}Y\right\|^2 - \left\|P_N Y\right\|^2}{\hat{\sigma}^2\, d(M^{(1)}-M_0^{(1)})},$$

distributed (assuming normality of Y) as

$$\mathcal{F}_{d(M^{(1)}-M_0^{(1)}),\, d(V^{(1)}-M^{(1)})+d(V^{(2)});\, \gamma},$$

where

$$\sigma^2\gamma^2 = \left\|P_{M^{(1)}-M_0^{(1)}}\mu^{(1)}\right\|^2 + \left\|P_{M^{(2)}-M_0^{(2)}}\mu^{(2)}\right\|^2 + \left\|P_{N_0}\mu\right\|^2.$$

What simplifications take place in these formulas when $M_0^{(2)} = M^{(2)}$?   ○

G. Consider an $I \times J$ two-way additive layout with one observation per cell, augmented by m replications for the IJ^{th} cell, that is, suppose

$$Y_{ij}, \quad i = 1, \ldots, I, \ j = 1, \ldots, J, \quad \text{and} \quad Y_{IJ\ell}, \quad \ell = 1, \ldots, m,$$

are uncorrelated random variables with common variance σ^2, and

$$E(Y_{ij}) = \nu + \alpha_i + \beta_j \quad \text{for } i = 1, \ldots, I, \ j = 1, \ldots, J$$
$$E(Y_{IJ\ell}) = \nu + \alpha_I + \beta_J \quad \text{for } \ell = 1, \ldots, m$$

with

$$\sum_{1 \le i \le I} \alpha_i = 0 = \sum_{1 \le j \le J} \beta_j.$$

Using the foregoing material wherever appropriate, work out explicit formulas for:

(1) the GMEs $\hat{\nu}$, $\hat{\alpha}_i$, and $\hat{\beta}_j$ of ν, α_i, and β_j;
(2) the variances and covariances of these GMEs;
(3) the usual unbiased estimator $\hat{\sigma}^2$ of σ^2;
(4) the F-statistic for testing

$$H\colon \alpha_i = 0 \text{ for each } i \quad \text{versus} \quad A\colon \alpha_i \ne 0 \text{ for some } i;$$

(5) the degrees of freedom and noncentrality parameter for the above statistic; and
(6) Scheffé's simultaneous confidence intervals for all linear combinations of the α_i's.   ○

H. What modifications would you make in the formulas of parts **A** through **F** if assumption (6.2) on the dimension of $M^{(2)}$ were dropped?   ○

REFERENCES

The references given here comprise only those works cited in the text, and papers from which the problem sets were derived.

[1] Baranchik, A. J. (1970). A family of minimax estimators of the mean of a multivariate normal distribution. *The Annals of Mathematical Statistics*, **41**, 642–645.

[2] Berger, J. O. (1976). Admissible minimax estimation of a multivariate normal mean with arbitrary quadratic loss. *The Annals of Statistics*, **4**, 223–226.

[3] Bickel, P. J., and Doksum, K. (1977). *Mathematical Statistics: Basic Ideas and Selected Topics.* Holden Day, San Francisco.

[4] Brown, L. D. (1971). Admissible estimators, recurrent diffusions, and insoluble boundary value problems. *The Annals of Mathematical Statistics*, **42**, 855–903.

[5] Bryant, P. (1984). Geometry, statistics, probability; variations on a common theme. *The American Statistician*, **38**, 34–48.

[6] Cleveland, W. S. (1971). Projecting with the wrong inner product and its application to regression with correlated errors and linear filtering of time series. *The Annals of Mathematical Statistics*, **42**, 616–624.

[7] Efron, B., and Morris, C. (1973). Stein's estimation rule and its competitors — An empirical Bayes approach. *Journal of the American Statistical Association*, **68**, 117–130.

[8] Efron, B., and Morris, C. (1976). Families of minimax estimators of the mean of a multivariate normal distribution. *The Annals of Statistics*, **4**, 11–21.

[9] Haberman, S. J. (1975). How much do Gauss-Markov and least squares estimates differ? — A coordinate-free approach. *The Annals of Statistics*, **3**, 982–990.

[10] Halmos, P. R. (1958). *Finite-Dimensional Vector Spaces*, second edition, Van Nostrand, Princeton, New Jersey

[11] Herr, D. G. (1980). On the history of the use of geometry in the general linear model. *The American Statistician*, **34**, 43–47.

[12] Kruskal, W. K. (1960). Untitled. Unpublished monograph on the coordinate-free approach to linear models.

[13] Kruskal, W. K. (1961). The coordinate-free approach to Gauss-Markov estimation, and its application to missing and extra observations. *Proceedings of the Fourth Berkeley Symposium of Mathematical Statistics and Probability*, 1, 435–451.

[14] Lehmann, E. L. (1959). *Testing Statistical Hypotheses.* Wiley, New York.

[15] Rao, C. R. (1965). *Linear Statistical Inference and Its Applications.* Wiley, New York.

[16] Saville, D. J., and Wood, G. R. (1986). A method for teaching statistics using *n*-dimensional geometry. *The American Statistician*, 40, 205–214.

[17] Stein, C. (1966). An approach to the recovery of inter-block information in balanced incomplete designs, in *Festschrift for J. Neyman,* 351–366, Wiley, New York.

[18] Stein, C. (1973). Estimation of the mean of a multivariate normal distribution. *Proceedings of the Prague Symposium on Asymptotic Statistics*, 345–381.

[19] Strawderman, W. E. (1971). Proper Bayes minimax estimators of the multivariate normal mean. *The Annals of Mathematical Statistics*, 42, 385–388.

[20] Wald, A. (1942). On the power function of the analysis of variance test. *The Annals of Mathematical Statistics*, 13, 434–439.

[21] Zabell, S. (1994). A conversation with William Kruskal. *Statistical Science*, 9, 285–303.

INDEX

Printed in the United States
By Bookmasters